Praise for Karen Dawn and *Thanking the Monkey*

"Vegetarianism was a source of Gandhi's successful nonviolence, and cruelty to animals is a predictor of everything from domestic violence to serial killing. With wisdom and insight, Karen Dawn introduces readers to the fact that animal rights are an organic part of social justice movements everywhere. Her book is a bridge between worlds for both the committed and the curious."

—Gloria Steinem

"I am sick of being ignorant! This book is education with a smile, information with a kiss from a dog who just drank out of a toilet, and should be required reading for all college students."   —Anthony Kiedis, Red Hot Chili Peppers

"In an age, and in a medium, in which people seem to be content with quick clichés and approximations, Karen Dawn writes a lucid and accurate prose that could be held up as a model to her profession. Her work treads that fine line between the provocative and the counterproductively undiplomatic assuredly."

—J. M. Coetzee, Nobel Prize–winning author of *Disgrace*

"Karen Dawn's work is powerful—compelling and well argued, intellectually sound and passionate."   —Paul Haggis, Oscar–winning writer-producer of *Crash*

"Karen Dawn's new book is a graceful invitation to open your heart to animals. With well-reasoned arguments and engaging, witty prose, Dawn goes beyond the usual categories of liberal and conservative with a message of compassion that speaks to everyone. The book is a fast read that can change your life forever."

—Matthew Scully, former senior speechwriter
for President George W. Bush and author of *Dominion*

"Charming and thoughtful, witty yet engaged, it would be a hard heart that could put this book down without having old prejudices seriously shaken. *Thanking the Monkey* will make you sit up and take notice and probably, too, drop your fork and reach for the fruit."

—Jeffrey Masson, bestselling author of *When Elephants Weep*
and *Dogs Never Lie About Love*

"*Thanking the Monkey* is a fun and funny look at a sad subject. As you laugh your way through this book it will change the way you view animals and have a deep impact on your life." —Tamar Geller, bestselling author of *The Loved Dog*

"The torture, the experimentation on animals is such a serious subject that a lot of people just want shut their eyes to it. That is why I am so glad Karen Dawn has written this new book. It's just like when we do Comic Relief—we deal with serious subject matter, but at the same time we are able to joke and bring humor to it. Get this book!" —Bob Zmuda, president and founder of Comic Relief

"I am so excited about *Thanking the Monkey* because I have loved Karen Dawn's approach to animal rights for a long time. I love taking it and making it fresh, making it young and making it interesting." —Persia White, star of *Girlfriends*

"Karen Dawn's book made me realize I know very little about animal rights, and I would like to know more. It is the kind of book you can pick up and get something of value out of without reading chapter 1 first and chapter 2 second. You can flip it open virtually anywhere and find something you didn't know." —Bruce Greenwood, from *John from Cincinnati*

"*Thanking the Monkey* is a fun way to talk about animal rights and the interconnectivity of humans and other beings. It will appeal to the younger generation and people who have a harsher idea about animal rights books." —Daniela Sea, from *The L Word*

"If you care about the planet, if you care about yourself, you must care about the animals. Please find out the truth about animal abuse for yourself. Karen Dawn's book is the best resource around to help you save the planet, and yourself." —John Feldmann, lead singer of Goldfinger and producer for the Used and Good Charlotte

Monty Marsh

About the Author

Born in the United States, KAREN DAWN grew up and studied in Australia. She pursued a science degree colloquially tagged "rats and stats," so she knows firsthand that people's views and habits can shift. She hopes her furry subjects will shine forgiving blessings upon this book.

She worked as a news researcher and writer for Australia's national nightly news show *The 7:30 Report*, then moved to New York, where she played the downtown music scene—and made fruit salad at the Saint Francis Xavier soup kitchen every Sunday. After reading *Animal Liberation* she was moved to devote her efforts to those most abused by society and least able to help themselves—the animals.

Karen founded the animal advocacy media watch DawnWatch.com. As a spokesperson for the animals rights movement she has appeared on MTV and hosted talk shows on major radio stations. Her opinion pieces have been published in many newspapers, including the *Los Angeles Times* and the *Washington Post*. This is her first book.

karen
Dawn

HARPER

NEW YORK ● LONDON ● TORONTO ● SYDNEY

Thanking the Monkey

the

Monkey

Rethinking the

Way We

Treat Animals

HARPER

THANKING THE MONKEY.

For information address HarperCollins Publishers,
10 East 53rd Street, New York, NY 10022.

HarperCollins books may be purchased for educational,
business, or sales promotional use. For information please write:
Special Markets Department, HarperCollins Publishers,
10 East 53rd Street, New York, NY 10022.

FIRST EDITION

Designed by Joy O'Meara

Library of Congress Cataloging-in-Publication Data is available upon request.

ISBN 978-0-06-135185-3

08 09 10 11 12   ID/RRD   10 9 8 7 6 5 4 3 2 1

In loving memory of David Bale, Gretchen Wyler, Uncle Harry, and James

contents

Contents

six
Animals Anonymous
Animal Testing
215

seven
The Greenies
273

Contents

Thanking the Monkey

chapter one

welcome to the world of animal rights

Welcome to the world of animal rights. When I tell people I work full-time as an animal rights activist, many of them have questions. Am I vegan? If so, why—aren't California's "Happy Cows" really happy? Do activists who target medical research like mice more than men? And who belongs in the zoo—are the animals there good ambassadors for their species? After spending eight years fully immersed in animal rights issues monitoring the media for DawnWatch.com, I decided I was ready to tackle those questions and wanted to do so in as friendly and fun a manner as possible—and so we have this book. In it, I hope to help dispel the myth that animal rights activism is radical and unreasonable. In fact, as you read of the cruelty we offer animals as thanks for what we take from them you may see radical departures from your own standards of reasonable decency.

Karen Dawn

animal rights vs. human rights

Let's start by addressing some common questions animal rights activists get asked.

Why worry about animal rights when there is so much human suffering in the world?

Animal rights activists are asked that constantly. And you wonder why we tend to be feisty! Why don't people ask human rights activists how they can do their work when there is so much animal suffering in the world? Seriously, though, part of the answer is in the question. Even somebody who does nothing to end human hunger wouldn't justify his apathy by telling relief workers that there are more important things to worry about.

That's because society as a whole acknowledges that human suffering matters. Animal suffering, however, is treated as trivial, even as billions of beings endure unimaginable institutionalized cruelty. To those touched by the suffering of animals, the injustice of the suggestion that animals just don't matter is a call to action.

The question, moreover, is based on a faulty premise. It suggests that compassion is like a pie we must divide into parts, and that if we offer big pieces to some, others will get left with slivers. But compassion is not some sort of finite substance that might run out. It is more like a habit we get better at as we practice, and the

I do ALL the heavy lifting, and you have the audacity to call this "fair trade" coffee?

BIZARRO.COM

Bizarro © Dan Piraro/King Features Syndicate

2

animals are a good place to start exercising it—for their sake and for ours. George Angell, the founder of the Massachusetts Society for the Prevention of Cruelty to Animals, put it well when asked why he focused on kindness to animals when there is so much cruelty to people in the world. He said, "I am working at the roots."[1]

Before I extended my own efforts toward animals, I worked every Sunday, for six years, in a soup kitchen for New York's homeless people. I worked alongside many fellow vegetarians. And when I saw the film *Amazing Grace*,[2] I was not surprised to learn that William Wilberforce, who led the parliamentary campaign against the British slave trade, was also one of the founders of the Society for the Prevention of Cruelty to Animals. That's because compassion and cruelty are not species-specific. Most of us have heard that serial killers usually start by killing animals. The same compulsion drives the killers' behavior when they move on to humans; the urge to hurt just becomes so strong that it outweighs societal norms and fears of legal retribution. So it is with less active cruelty, with the closing of our hearts that has us sit by as others suffer. The compassion shutdown switch that allows us to chew pieces of veal while blocking out thoughts of baby calves alone in crates is the same switch that guides us to change TV channels away from news of children starving in Darfur. We don't want the images to hamper the taste of the meat or our enjoyment of the wine we are drinking, a bottle of which costs more than it costs to feed a child in Darfur for a month. When we disengage that switch, when we get out of the habit of closing our hearts, the world will be better for the calves and the kids.

Animal what?

What exactly do you mean by animal rights?

Funny you should ask—I will surely be challenged to a duel or two over the heading of this chapter, for I use the term "animal rights" more loosely than some would like. I use it to refer to what is commonly known as the animal rights movement—those who devote themselves to advancing the interests of animals and who discourage the use of animals as objects of commerce. For some activists the term "animal rights" is

*"I'd like to hear less talk about animal rights
and more talk about animal responsibilities."*

literal; those activists seek legal rights for members of other species. Though they do not wish to earn nonhuman animals the right to vote—any more than they wish to see that right given to human children—they do wish to see animals granted the right, as it is put by the animal rights lawyer Steve Wise, to "bodily liberty and bodily integrity."[3] That means no cages, no knives, and no scalpels.

Political conservatives in our movement generally hold that animals don't have rights at all, but that we have responsibilities toward them. One of the leading proponents of that view is Matthew Scully, who was a senior speechwriter for President George W. Bush. He argues that our basic responsibility to other animals is to treat them with mercy.[4] Scully is now vegan, which means he believes his responsibility to animals includes abstaining from eating them or the products of their common abuse, while living in a society with so many other alternatives.[5] If he were to persuade the world to follow his lead, would it matter, at least to the animals, whether or not he spoke about rights?

The Anti-Welfare Warriors

Some activists seek total universal animal liberation and disdain welfare campaigns that attempt to ease the suffering of animals in captivity. Others don't believe that the utopian animal liberation vision will be realized any time in the near future, and feel driven to work on campaigns that help relieve some of the worst animal suffering endured today. Indeed, there are rifts in our movement over whether the fight for animal rights can also include efforts to improve animals' welfare. Some animal rights activists, who often call themselves abolitionists, feel that animal welfare laws ultimately work against the animals, weakening our case for animal liberation. Those activists might suggest, for example, that it is easier to persuade people to stop eating veal while calves are kept in crates and deprived of iron. They argue that welfare improvements just allow people to keep eating animals and alleviate the guilt that would eventually make people abstain.

© Anthony Freda

Even if that were so, can you imagine Amnesty International campaigning against laws that forbid the torture of political prisoners because the prisoners shouldn't be in jail at all, and because their case will be stronger if the torture continues?[6] Assuming that animal rights activists would attempt to negotiate the release of any imprisoned colleagues, and would also request warm beds and nourishing vegan food for them, how can we refuse such consideration to the nonhumans we volunteer to represent? It has been argued, persuasively in my opinion, that such a stance reduces the animals to objects in service to abolitionist ideology.[7]

If we look at the history of social justice movements we see that improving conditions for the oppressed has not hampered the fight for liberation. While women as a group have yet to earn equal pay for equal work, surely nobody would suggest that granting women the right to vote obstructed the road to eventual societal equality. Laws forbidding the beating of slaves came before, not instead of, laws precluding slavery in the northern states, and were part of the movement that led to emancipation throughout the United States. That's because, as Robert Cialdini explains, people tend to make consistent choices.[8] The consistency theory is the basis for all foot-in-the-door sales techniques, and for animal enterprise's "slippery-slope" arguments against granting any animal welfare reforms. When society supports welfare measures aimed at ending some of the most hideous industrial abuses of animals, it acknowledges that animals matter. Consistent with that position, those who have supported those changes are more likely than others to ponder their personal use of animal products.

That theory was beautifully exemplified in Elizabeth Devita-Raeburn's forthright article "An Ambivalent Vegetarian," in *Self* magazine.[9] She made it clear that learning about slaughterhouse reforms did not ease her conscience when she craved meat. She wrote of those reforms: "But the need for them made me feel even worse. Clearly these are not dumb, insensible creatures who are oblivious to whether they live or die. Quite the reverse."

People in the food production industries (who have a strong record of knowing how to shape public behavior) do not swallow the argument that welfarism hampers abolitionism. An editorial in *Feedstuffs* warned readers that the food industry is losing the battle against animal activists. The piece listed numerous successful welfare campaigns and then proclaimed, "It's about raising animals for food and the activists' agenda is to end that practice. It will take decades, but they are the ones who are winning—piece by piece by piece."[10]

I have no desire to hide my agenda, and am happy to admit that I think humans are evolving toward vegetarianism. I form that hypothesis partly from noting the high proportion of vegetarians among history's greatest thinkers—apparently significantly higher than the few percent estimated in the general population. Pythagoras, Plutarch, Leonardo Da Vinci, Tolstoy, Twain, Bernard Shaw, Kafka, Einstein, and Gandhi come to mind. I see welfare reforms as steps on the way—part of that evolution. Now that doesn't mean this book is not for you if you don't particularly want to see yourself or the world go veggie! Any changes made by anybody can help make the world a more compassionate place, and public support for welfare reforms makes a huge positive difference.

Whatever doubt I had about that position dissipated when I saw the documentary *Beyond Closed Doors*.[11] This moderately toned exploration of factory farming includes video of sows in gestation crates. That image is salient for me, as I credit my first awakening of interest in the animal rights movement to a brochure from the Humane Farming Association that displayed photos of sows housed in those crates. The huge, intelligent animals cannot even turn around or lie down with legs outstretched. They spend their lives staring ahead, biting on the bars in front of them, and going mad. *Beyond Closed Doors* also shows contrasting video of sows running in the fields of Niman Ranch, a farm certified by welfare groups as a humane meat producer. Sadly, humane labeling can be misleading. In chapter 5 of this book I will note that some of the humane labels are absolute scams, and I will discuss abusive practices even on farms certified by animal welfare groups. Yet one cannot look at sows in individual crates, and sows running together across fields, and say it really isn't worth letting those sows destined for dinner tables at least really live while they are alive—in fields rather than crates.

I was struck in that film by Dr. John Webster's comment, "I do point out that no sentient animal has the right to escape death."[12] Some of the anti–welfare–reform arguments seem to ignore that truth. While we can choose not to cause an animal's early death, we cannot save any animal from mortality, any more than we can save ourselves. We can, however, work to save animals from horrendous suffering. Many animal rights activists who would prefer a decent life—even truncated—to a life of pure suffering, feel it is incumbent upon us to work to ease the suffering of those animals who will have their lives cut short for the human food chain. Such efforts may not give us the seductive satisfaction of influencing the behavior of other people as much we would like, and they do not fit into our view of a perfect world, but they make a huge difference in the lives of the animals. That's why many vegans work hard on welfare campaigns and why this animal rights book includes welfare issues.

People for the Ethical Treatment of Animals (PETA), the world's best-known animal rights group, does the rights-welfare balancing act, aiming to be both visionary and pragmatic. PETA works on welfare campaigns but has "abolitionist" language in its credo. PETA says, "Animals are not ours to use—for food, clothing, entertainment, experimentation, or any other reason."[13]

Not all compassionate people will automatically agree. For example, if we give some hens a good life on spacious grounds, protected from predators, might we not have a fair trade if we use their eggs? Probably not—I will explain why later in this book. Yet I worry that arguing about the right ever to eat animal products distracts people from learning that when they buy eggs, they generally support shocking cruelty that most people, not just animal rights activists, would hate to condone. We usually thank the hens for their life-sustaining gifts by giving them lives worse than death.

Karen Dawn

One of my favorite definitions of "animal rights" comes from the late, great Gretchen Wyler, who founded the Genesis Awards, a kind of animal-friendly Oscars, now run by the Humane Society of the United States (HSUS). Without defining her vision as either animal welfare or animal rights, she simply told us, "Animals should have the right to run if they have legs, swim if they have fins, and fly if they have wings."[14] It is a vision that to most compassionate people seems fundamentally and obviously, well, right. Yet it is far from reality for billions of animals alive today.

Here's animal-friendly actor Christian Bale at the Genesis Awards. Christian's dad was on the Genesis Board until he passed away in 2003.

Photographs courtesy of HSUS Hollywood Office.

A golden-age Broadway star of *Damn Yankees* and other hits, Gretchen Wyler founded the Genesis Awards in 1986 when she held a small luncheon to thank a few animal-friendly media people. Now approximately one thousand members of the media and the animal protection community attend the televised, celebrity-studded annual celebration in Beverly Hills. © Long Photography, Inc.

spotlight on speciesism

Don't animal rights activists prefer animals to people?

We get asked variants of that question—like whether we would save a dog or a person if they were both drowning. Well . . . what person? Think of your least favorite world figure. Who would you save if he and your dog were both drowning? I bet you can think of somebody who might come in second to your dog's favorite chew toy.

While we are on those nasty public figures: Sometimes people ask, "If kindness to animals and humans is connected, then why was Hitler a vegetarian?" He wasn't. Hitler wrote about the gullibility of the masses, suggesting that the bigger the lie, the easier it is to make the public swallow, because nobody expects people to "fabricate colossal untruths."[15] Apparently people who wanted to discredit the idea of ethical vegetarianism took his words to heart and started spreading a lie that got so big it became folklore, even ending up in a movie review in the *New York Times*. That paper, which takes pride in its record, printed a retraction that stated, "A film review about 'Downfall,' which looks at Hitler's final days, referred incorrectly to his diet. Although the movie portrays him as vegetarian,

The Evolution of Man

9

he did eat at least some meat."[16] While Hitler's doctor had recommended a vegetarian diet for him, his passion for pork sausages made it impossible for him to follow that advice.

Why does refuting that myth matter? Because Hitler is so often used as the archetype of everything evil, his supposed vegetarianism was the perfect piece of proof that those who care about animal rights don't care about human rights. Rubbish. Thanks to the *New York Times* for shooting down a lie that has been flying high for too long. Can we bury it now?

On Equating Life

Do animal rights activists equate human and other animal life?

I love the answer given by Professor Gary Francione when a reporter asked him what seemed to be a variant of that question. Asked whether he would kill a rat with bubonic plague who was trying to get into his house, Francione answered, "I would kill you if you had bubonic plague and were trying to get into my house!"[17] His answer points to the inadequacy of that type of question—I don't think any of us will ever find ourselves in situations where other things are

© Anthony Freda

equal and we really have to make the existential choice between human and animal life. I hope to convince you in this book that neither is that choice being regularly made by others, by members of the biomedical industry, for example. They might want you to think you should thank the monkey for giving her life to save yours, but she is just as likely to have died, unwillingly, to help you clean your kitchen or clear up your toe fungus in two days instead of three.

There actually are people who say they equate animal life with human life—even their own. I have a friend in the animal rights movement who will eat rabbit food all day, but never ever eat a rabbit. He tells me that if he were starving on a desert island he would not kill a rabbit, because the rabbit has as much right to life as he does. And then I know a man in the antiabortion movement who would forfeit the life of his wife if she were destined to die in childbirth rather than abort a three-day-old zygote. But such rigid views, which make it easy to write off those on the other side of an issue, are hardly typical of those in either movement.

Peter Singer is often called the founding father of the modern-day animal rights movement (though he doesn't argue for rights). His groundbreaking book *Animal Liberation* brought me and brought many of the animal rights movement's current leaders to the cause. Singer tells us that we should avoid speciesism, which is the preference for individuals based entirely on their membership in a certain species—ours.[18] His views are often misrepresented, as it is suggested that he values human and other life equally. But in *Practical Ethics*[19] he actually argues that it is okay to value the lives of normally developed conscious humans over those of other animals, if we base the value not on species but on certain attributes. And it just so happens that those definitive attributes, according to his litmus test, are the ones that normal humans have, and that, he believes, other animals don't—such as the ability to make plans for the future. Does that mean human lives matter most because we are the only ones making plans to shop at Wal-Mart, or to go watch the game at Hooters? What if we are making plans to annihilate the world? Even if our efforts were stumped, and we were locked up and could hurt nobody, is the life of a human who wishes to annihilate the world really worth more than the life of, say, gentle Koko the gorilla—or of your beloved dog?

Mutts © Patrick McDonnell/King Features Syndicate

Saying it is not speciesist to value most human life more than other life because of attributes apparent in normal humans is like saying you aren't racist if you value whites more than blacks, as long as you equally value those of the Negroid race whose skin is so light that they look white. Your judgment isn't based on their race, it is based on the color that just happens to usually go along with the race. Your argument might be technically correct, but I don't think many people would buy it. I think I would call you racist.

I question judgments that are based on our performance on some universal ethical test for the worth of a life, when the test has been designed using, and is also testing for, characteristics that happily seem to pertain most strongly to members of our own species. Why is the test not based on the gentlest nature, or the ability to live on Earth without destroying it? Perhaps because we are not going to design a test we are destined to flunk.

Talk of incomparable human talents reminds me of Rush Limbaugh, who has laughed at people who think dolphins might be similarly gifted. He wondered aloud why dolphins haven't built automobiles and highways. And he—or was it another one of those insightful commentators—noted that dolphins haven't written Mozart's concertos.[20] I don't think Rush has built any highways and when I turn on those shows I generally don't hear Mozart. The only concerto I hear is the Rush toilet flush; I think it's in B minor. Anyway, I can't imagine why dolphins would need highways. And I wouldn't know if dolphin sonar sounds to dolphin ears something like Mozart. But I am impressed by dolphin gifts that I don't share, such as their astounding navigational ability. I also find it interesting that dolphins have larger brains than humans—as do whales and elephants. Scientists try to get around that by suggesting that intelligence is linked not to overall brain size but to the brain-to-body mass ratio, which has humans come out just slightly ahead of dolphins—and, oops, way behind hummingbirds.[21]

There is no way to look at dolphins and know whether they or we humans are worth more in some universal sense. We simply, naturally, are inclined to care most about our own. But as the saying goes, "We buy with our hearts and justify with our minds." Some feel we need to justify our instinct with value judgments about the worth of the other lives in question.

© Paul Conroy/Joss Stone

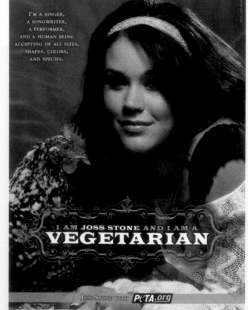

I'M A SINGER, A SONGWRITER, A PERFORMER, AND A HUMAN BEING ACCEPTING OF ALL SIZES, SHAPES, COLORS, AND SPECIES.

I AM JOSS STONE AND I AM A **VEGETARIAN**

JOSS STONE WITH PeTA.org

Courtesy of PETA and Joss Stone

"I'm a singer, a songwriter, a performer, and a human being accepting of all sizes, shapes, colors, and species" (**Joss Stone**).

OUTLAND   **BY BERKELEY BREATHED**

DOG APPRECIATION DAYTIME "NAKED IN THE PERIWINKLE"

THERE! *THAT* IS WHY MAN IS THE MORAL SUPERIOR TO ANIMALS!

OBLIVIOUS TO ANY NOTION OF "THE FUTURE", THE BEAST DUMBLY DOZES HIS LIFE AWAY NAKED IN THE PERIWINKLE!

HEY! GET A JOB! GET A MORTGAGE! GET SOME BONDS! GET A RELIGION! GET A MARRIAGE! GET A DIVORCE! GET A LEXUS! GET GOIN' ON THAT INFO HIGHWAY!

FOR CRYIN OUT LOUD, YER GONNA CROAK! HURRY UP AND GET SOME STUFF!!

ZZZ... SNRK.

© 1989, 1994 by Berkeley Breathed. Distributed by The Washington Post Writers Group. Reprinted with Permission.

Mirror, Mirror

Mirror, mirror, on the wall, who's the most like us of all?

That's what is asked by the "mirror test," one of the tools humans traditionally use to justify speciesism. The pinnacle is the x test: While an animal is asleep, some sort of mark is put on her head or body. Then the animal is put in front of a mirror and the researcher observes her reaction.

Most animals show no interest in the x. Scientists cite an animal's indifference as proof that she doesn't realize that she is looking at an image of herself. Poor performances on x tests are even offered as proof that animals have no understanding of themselves as distinct beings. That presumed lack of self-awareness is supposed to mean that their lives matter less.

Dolphins have passed the mirror x test, and so have chimps. In 2006, elephants were added to the list. One elephant at the Bronx Zoo, Happy, looked in a mirror and repeatedly touched the tip of her trunk to an x that had been painted on her cheek. That feat led to several newspaper reports asking whether, since elephants are self-aware like us, it is appropriate to hold them captive in small pens for our entertainment. Apparently the day before Happy showed interest in an x on her face, that had not been an ethical concern.

Interestingly, though the other elephants demonstrated clear understanding that they were in front of mirrors—for example, one used the tip of her trunk to pull back her lips and get a better look inside her mouth—they didn't show interest in the marks painted on their faces. If Happy had not been part of the trio used in that x mirror test, elephants would have failed it. Wouldn't that suggest the x test should be taken with an elephant-size mountain of salt?

Dogs fail the x test. Some researchers have suggested, therefore, that they don't understand mirrors. My bedroom closets form a wall of mirrors. If I talk to Buster, my big, beautiful elderly mutt, while he is lying facing in their general direction, he will often look up and stare at me in the mirror instead of turning around to face me— exactly as a person would. While Buster is deliciously affectionate and a magnificent singer, I don't think he is a genius. I doubt that his mirror response is unique.

© Doug Hall/PCRM

If your dogs didn't understand that they saw themselves in mirrors, wouldn't they always be barking at the other dogs they perceive in any room that has mirrors at doggie eye-level?

It makes perfect sense, however, that dogs fail mirror tests that use marks on their faces, or might fail any tests based entirely on visual sense. Though dogs seem to have good night vision, compared to their other senses their eyesight sucks. I am told that when I am not with my dogs they get excited when any woman of my height, build, and hairstyle approaches, until she is close enough to smell—which for them seems to be at about a hundred feet (give or take a few yards, depending on how often she showers). The canine sense of smell is estimated to be about 250 times stronger than ours. Dogs navigate the world with their noses and ears, and perhaps with senses we know nothing about. So how ridiculous is it of us to decide that their behavior in front of a mirror, an instrument that relies on visual cues, tells us anything about their self-awareness?

I would like to try the following test: While a dog and a person are asleep, let's put a microscopic drop of a pungent substance, either delicious or noxious, on their forearms. When they wake up, let's see whether the dog or the person reacts to it and scratches or licks at it. My money is on the dog. Most people are only olfactorily attuned enough to notice the smell of fresh popcorn in a movie theater—particulary if they are just waking up. But a person wouldn't notice a miscroscopic drop of anything. The dog, however,

would be sure to notice and attend to any tiny drop with a surprising smell. Would that then prove that dogs are self-aware and people are not?

In any test we humans design we can prove that other animals are not exactly like us, but we are likely to misinterpret the differences—assuming, for example, that a lack of vain interest in mirrors means a lack of understanding of oneself as an individual.

Why do we devise such tests? Because we don't want to admit that we are simply speciesist. It is our instinct to care most about our own kind (perhaps because it ensures the survival of our species), and we prefer to find excuses to justify that instinct rather than acknowledge it as another realm of xenophobia. I am not suggesting that our instinct is bad, and that we must equate other animals with humans in our society. But I don't think it is useful to try to prove that we are fundamentally more worthy than they are. With all biases, if we admit them and stop trying to justify them, we will be in a better position to make well-reasoned and compassionate decisions about the extent to which we wish to support or to work against them.

Karen Dawn

Are There Limits?

Even if we agree that it is natural and not wrong to care most for our own, to what extent do we take that?

Would I trade the life of my neighbor's child for the life of my own child? The government wouldn't let me. But the law to stop me exists because there is a basic understanding that if I had total power (as we humans do over other animals) I probably would make that trade—almost anybody would.

Would I trade the life of my neighbor's child for my child's eyesight if I could? I doubt it, but maybe. Would I trade his life for my child's leg? I don't think so. For my child's kidney? No. How about for my child's appendix, to spare my child the pain of the operation? You've got to be joking! Nobody would.

Eventually we come to a place where we see that the preference for our own has some moral limits on which everybody agrees. So while acknowledging that we, as a species, have the power and desire to place our lives above those of other animals, why not ask ourselves about the moral limits that must temper our preference for our own?

Those who make money from biomedical experimentation would love people to believe that animal rights activists prefer rats to kids. (We can laugh off that idea and be relieved they don't put us on the spot by asking if we prefer rats to vivisectors.) If animal experimentation were all about painless tests on rats to cure cancer, there wouldn't be an antivivisection movement with millions of supporters. You would be hard-pressed to find an animal rights activist who would not kill a rat to save a child. You would do it in a second, right? Would you kill a dog to save a child? I bet you would. Would you kill thousands of them so that a child can have a new cold medicine that lasts for eight hours rather than six hours? Maybe not. How about for a cold medicine that does exactly the same thing as other currently available drugs from which the drug company no longer profits because the patents have expired? Of course not! But that is what drug companies do, and we buy their products.

can't we all just get along?

Somewhere in the midst of the examples above we crossed the line into an arena where we find a lot of agreement.

And that is what I hope to do with this book—not to fight with my readers and win the battle for animal rights, and not to force my values on others. The idea, rather, is to tell you everything you wanted to know about animal rights but were afraid to get into a fight about, and to let you weigh that information against your own values. You can decide what practices you find acceptable or not, and how you might avoid supporting what you cannot condone. Who knows? You might want to go vegan! Or, for now, you might find you just want to change where you buy your meat. Perhaps you'll decide to take your kids to the puppet show instead of the circus. There are so many ways one can make a difference for the animals—or, as some of us playfully call it, to start "thanking the monkey." If you are not immediately spurred to make any changes—if you aren't ready to do any personal monkey thanking—perhaps you'll just do some thinking about how society treats animals. You might come to see the fight for animal rights as a worthy stance against some serious and surprisingly far-reaching wrongs.

© Anthony Freda

chapter two

slaves to love: pets

canine americans?

Wayne Pacelle, the president of the Humane Society of the United States, opened his address at a Taking Action for Animals conference[1] with the suggestion that we should perhaps stop referring to dogs as dogs, since that word has derogatory connotations, and should instead start calling them Canine Americans. He got a good laugh from the audience and an astounding response from those against our movement who had come to the conference to monitor animal rights activists. They sent out press releases suggesting Pacelle was serious! Sorry, folks—nobody is seriously suggesting we start calling dogs Canine Americans—though the IRS is hopeful and has already drafted a regulation on their tax responsibility. Pacelle's joke did, however, brush against some truth. Just as other social movements have discarded demeaning terms, we see calls among animal rights advocates for some changes in language.

The animal rights community dislikes the term "owner"—the idea that one can own another being. Animal advocates generally adopt rather than buy animals, and we see ourselves as guardians. I use the term "owner" only when discussing a case of abuse, where it is clear that the person has treated the animal as an object that is owned rather than as a sentient being who is cared for.

Many activists dislike the term "pet" and prefer "companion animal." Dictionary definitions of the word "pet," however, refer to one who is pampered and treated with

special care and affection. Some may see the term as demeaning, similar to referring to our animals as our babies, arguably inappropriate since they are fully grown members of other species. But until dogs start driving themselves to the vet we might have to accept that in caring for members of other species within human society, we are unavoidably in a superior role. That fact will not change even if we employ language that attempts to deny it. Why not, then, use language that accepts our role but makes clear that our animals are not objects or property to be owned, but pets, rather like children, to be pampered, adored, and, wherever possible, respected.

Animal advocates have many other issues of concern regarding language. For example, throughout this book from time to time I will refer to "humans and animals" when, of course, humans are animals. I have chosen accessibility over purity and hope those who prefer a more pure approach can bear with that choice.

should we have pets at all?

Mutts © Patrick McDonnell/King Features Syndicate

The strict animal rights position might be that we owe it to other species to leave them alone and let them live wild. That position, however, is not common in our movement. When one visits the PETA offices one has to walk carefully so as not to stumble over all of the animals who have been brought to work by their loving human caretakers. Most animal advocates decry pet overpopulation, which leads to the extermination of many millions of companion animals every year, but they do not object to the keeping of dogs and cats as pets.

Keeping dogs and cats is ethically different from keeping other animals. If an animal can do well away from us, in the wild, or in a truly wildlike habitat that we provide, then to lock him up and make him dependent on us is like putting him in jail for no crime. A domestic dog, however, cannot live in his natural habitat outside human society. His natural habitat is human society; he has been bred over thousands of years to live in it. Most domestic dogs do not have hunting skills, or coats to protect them from the elements. Many have features that would likely get them killed in the wild—like a basset hound's ears, which drag on the ground and would get tangled in the brush. Or a poodle's haircut would get him laughed out of the forest.

There is an argument for discouraging domestic "breeds." (See "Breedism" below.) Since nondomestic breeds, such as wolves and coyotes, can survive without us, as can feral cats, one could argue that in utopia they would live near us rather than with us, hanging around and befriending humans to scavenge cast-off food. That is probably the way the close human-canine relationship began. The animals would be free—like the squirrels I hand-feed on my back patio who come and go as they please.

Mrs. Squirrel knocking at our back door.

© Karen Dawn

Currently millions of companion animals bred under human supervision are exterminated for lack of homes. Millions of others live pathetic lives, kept alone in backyards day after day. That amounts to solitary confinement, considered cruel punishment for a human, but even crueler for a pack animal. Some are kept on chains, unsocialized, which is cruel to the animal and sometimes cruel to people he encounters.

Chained dogs are almost three times more likely to bite.[2] (See DogsDeserveBetter.com for more on that.)

We do see situations, however, where humans and other animals live together to the benefit of the animal companions. They might be in households with many animals who get lots of human attention. Or an animal can clearly be happy as the sole companion of a person who takes him everywhere, forming an interspecies pack of two. While, for the sake of the animals, the way we deal with them must change, phasing out the practice of keeping pets is not in their interest if we can, instead, guarantee them pleasant lives. We must make sure those animals we choose to breed are raised in a manner that takes into account all of their physical and emotional needs.

Breedism

Many breeders are not keen on animal rights folk. That's because we raise the possibility of putting them at least temporarily out of business. In fact, some animal rights activists would like to put breeders permanently out of business, arguing that it is not ethical to genetically manipulate animals for traits we find pleasing, some of those traits being detrimental to the animals. Think of the squish-faced pugs whose noses are so short they have a hard time breathing. Awwww, so cute! Europe is in the process of passing laws against breeding detrimental traits—including aggression—into dogs. Perhaps America will eventually follow suit, in which case the American Kennel Club won't be happy. A *New York Times Magazine* article on designer dogs reported that pugs have a narrowed birth canal, and therefore "like many breeds, virtually all pugs must be delivered by c-section."[3] That suggests that if a law were to forbid detrimental traits it would outlaw "many breeds." The same article told us, "Recently geneticists discovered that the mutation contributing to widespread deafness in Dalmatians is the same mutation that creates its signature spots." Therefore truly outlawing the breeding of detrimental traits would mean no more Dalmatians. Many other breeds would be banned for other reasons. It will be interesting to see how faithfully Europe follows through on the proposed laws.

The discussion of trait breeding, however, seems of lesser ethical importance than the basic discussion of reproduction—bringing new dogs and cats into human society as we kill millions of others. As healthy, friendly, loving, and lovable older dogs and cats long for shelter, and for life itself, many animal rights activists hold that there is no such thing as a "responsible breeder" who keeps churning out cute little puppies and kittens. Indeed, it is fun to have a new puppy or kitten in the house, but does the desire for that pleasure make it right to sentence another loving animal to die? And if we can save animals but instead choose to support those who breed them, aren't we sentencing homeless animals to death? We may think of humane euthanasia needles and forget how awful that sentence is—until we read a quote in the *New York Times Magazine* from somebody who works in the shelter system. Diane Mollaghan tells us the animals sense what is happening to them. She says, "A lot of them start to vomit or soil themselves the minute they enter the euthanasia room."[4]

Many animal rights activists hold that only after we have the overpopulation crisis under control should we encourage breeding, and then it should be regulated, never allowing animals bred to outnumber available homes.

The joy of mutts: a kiss from Buster Dawn.

© Monty Marsh

Karen Dawn

paritney's pet store adventures

People dying in Darfur just can't seem to compete for America's media attention with the adventures of Paris Hilton and Britney Spears. I think I will grab a "Best Week Ever" theme and just call them Paritney Spilton, because even though Britney's background seems more Kentwood Motel 6 than Paris Hilton, in recent times their public images and behavior have morphed. Paritney's tricks have ranged from silly to sad, one double act being of particular concern to animal advocates. No, not that one—when they let their little personal pets out for some air, California's climate was warm enough that we didn't protest the shearing. The events of true concern were Paritney's pet store purchases. First Britney strolled into a pet store in Beverly Hills and walked out minutes later with a puppy.[5] Paris did the same thing, at the same store, a couple of weeks after Britney.[6] If they'd spent those few minutes at the Victoria's Secret next door buying panties we might celebrate. But an impulse purchase of a companion for the next fifteen years? Well, I guess nobody expects them to have the dogs much longer than fifteen months, let alone fifteen years. After all, the most common age of dogs in shelters is around eighteen months old, puppies dumped when they outgrow their cuteness—and adoptability. Paritney seems unlikely to break that mold. We've already seen Paris go through dogs faster than mojitos, and I bet nobody warned Britney that California might soon force drivers to keep dogs safely restrained in vehicles—like babies.

While I fear for the long-term welfare of their dogs, my larger concern is that these highly respected public figures—well, okay, these unlikely role models for many kids—have publicly patronized, and thereby advertised, pet stores. Though the current pet overpopulation crisis makes all breeding questionable, of course there are different standards of breeders. At the bottom of the barrel are the puppy mill breeders, who produce the dogs sold in pet stores. The animals that emerge from puppy mills are overbred and sickly. Their mothers spend loveless lives in tiny, stacked cages, from which they are never released to eat, play, or even defecate. When their breeding days are over they are killed or sold to vivisection laboratories.

"Almost all of the puppies for sale at your local mall were churned out by breeding facilities called puppy mills. . . . The conditions common to most mills are so inhumane, the dogs are left in various states of physical and mental anguish.

"As long as there's a market for their babies, dogs will continue to be bred, neglected, and killed at puppy mills. . . . If you adopt an animal from the shelter, you'll not only save her life, you'll also help end the suffering of countless dogs at puppy mills" **(Charlize Theron)**.

From *Charlize Theron's Puppy Mill Expose*, PETA video available online at www.petatv.com/tvpopup/Prefs. asp?video=Charlize-Theron-pupply-mill.

When NBC's *Dateline* did an exposé on the link between pet stores and puppy mills, owners and employees at chain stores like Petland, and at high-end pet boutiques in fancy Manhattan neighborhoods, assured customers and *Dateline* that their dogs came from the very best breeders. Apparently those people had never before seen an investigative news show; in lying to the cameras they made public twits of themselves. *Dateline* traced the puppies back to horrendous mass breeding facilities in the Midwest.[7]

If you are one of those kids who emulates Paritney, throw away all your knickers if you must, but stay away from pet stores if you care about animals.

copycat killings

Pet stores might be the bottom of the barrel, but if the barrel had a basement it would be cloning. As millions of animals die every year for lack of homes, some of the ethical implications of cloning are obvious—similar to buying a puppy from a breeder, thereby encouraging those who contribute to the pet overpopulation crisis, when you could choose to support a rescue group and save a life. But there is a dark side that is less obvious: A front-page *Washington Post* story reported, "Though clones that survive to adulthood typically seem healthy, they die in inordinate numbers in the womb or just after birth, and the pregnancies appear to be stressful for the surrogate mothers."[8]

Bizarro © Dan Piraro/King Features Syndicate

"Stressful" is an understatement. It can take more than two hundred of the surgically impregnated surrogate mothers to yield a single clone birth. Those two hundred mothers are housed in laboratories and subjected to multiple invasive surgeries. The short lives of the vast majority of the newborn clones are painful. So much suffering—just so somebody can have an animal who looks, but can never be, just like one who died.

It looked for a while, with the launch of Genetic Savings & Clone in California, as if pet cloning were going to become big business. On September 26, 2006, however, that company sent a letter to its customers informing them that it was closing, as it had been "unable to develop the technology to the point that cloning pets is commercially viable." Capitalism at its best!

Expensive Mutts

The advent of designer mutts is heartbreaking for animal advocates. If you want a mutt, go to the pound. As the folks at AnimalRightsStuff.com say, put that environmental consciousness to work and recycle something that really matters.

If you like your shoe collection, make sure you adopt an older dog, because dogs teethe until about two years of age. I was lucky that the adoption of my dogs, who were both less than a year old, coincided with my going vegan. I had a designer shoe collection reminiscent of *Sex and the City*'s Carrie Bradshaw, which would have made it hard to give up wearing leather. My dogs made it easy—they thought the shoe rack was a snack rack.

If you have thought of adopting an animal but don't think you have enough time or enough room in your home to give her a great life, why not go to your local pound, find out who is scheduled to be killed tomorrow, and take two. They will keep each other company, and whatever life you can give them will be better than the death needles they faced. If you travel a lot, adopt older Chihuahua mixes. You can put them under your airplane seat, and almost all of the big fancy hotel chains (Starwood, for example) and the cheap motels (like Motel 6) accept dogs under twenty pounds.

Ritzy Rescues

Perhaps Paritney went to a pet store because she just had to have a Peruvian Powder-Puff Poodlehuahua, or some other trendy new brced. For others, it might not be a fashion issue—we might have grown up with animals of a particular breed and developed a soft spot for them. No worries—anybody in dog rescue will tell you that every breed has rescue groups. A search through Petfinder.com or a quick Google of any breed and the word "rescue" will lead you to a compassionate source. Now, if Peruvian Powder-Puff Poodlehuahuas were in particularly short supply, Paritney might have had to take a few hours off her busy party schedule to do some research and might even have had to wait a short time for the perfect dog. But surely it is worth a wait to save a life! And sadly, few breeds have any rescue waiting list at all. They are waiting right now on death row.

Paula and the Pit Bullies

I have a soft spot for pit bulls. It is hard not to be soft on them once you've had one. Sadly, they get a bad rap because they make sensational media villains. Their jaws are so strong that when they bite, they do horrible, even fatal damage—which makes great press. But when that French woman got a face transplant after "her dog" tore her old face off, the headlines didn't bother to mention that the dog was a Labrador retriever.[9] Imagine the headlines if he had been a pit bull! But he couldn't have been—most of France had already banned pit bulls. How ironic. That bit of breed-specific legislation certainly didn't help Madame Face Nouveau.

Left to their own devices, pit bulls are actu-

Paula Pitbull © Hakim Holloway

ally less likely than other dog breeds to attack humans. They are bred to fight dogs, with humans supervising the fights. If a human goes in to break up a fight and gets bitten, the offending dog is shot. Survival of the friendliest. In tests given by the American Temperament Test Society pit bulls consistently do better than many other breeds—including beagles and border collies.[10] Not only do they bite proportionately less often than many other dog breeds, we see a much smaller percentage of fatal attacks by pit bulls than by other large breeds, when the statistics take into account the far greater number of pit bulls in the population.[11]

Ridiculously affectionate, they are known as "leaners" because a pit bull will never sit next to you, they always lean against you—they have to be touching. The first night I had Paula Pitbull home from the shelter, she jumped up on my bed and curled up under my right arm. She has slept snuggled up against me like that almost every night for the last ten years. But her affection isn't only for me. When other people bend down to pet her, I have to warn them about the dangers of my pit bull: "Don't get too close—she'll stick her tongue in."

I am a big fan of the Web site BadRap.org, which smashes myths about pit bulls—myths suggesting they are devils or that they are angels. They aren't either. While there are exceptions, most pit bulls are dog-aggressive. BadRap gives you the lowdown on that, and even warns you not to leave your pit bull unsupervised with your other dogs. It also shares the beauty of the loyal and affectionate animals, and will give you tips if you want to adopt one.

While I know pit bulls can come with problems, I certainly don't support banning them and taking loving, loved, and altered dogs out of loving homes! Given my affection for pit bulls, however, people are often surprised to learn that I do support breeding bans on them, and even testified in favor of breed-specific spay-neuter legislation in a California Senate committee hearing. If you have read these pages in consecutive order, by now you know I would be happy to see a ban on almost all dog breeding until we have stopped killing dogs for lack of homes. But at the moment, pit bulls need those bans the most. The shelters are full of them and those darling dogs are being killed by the millions every year.

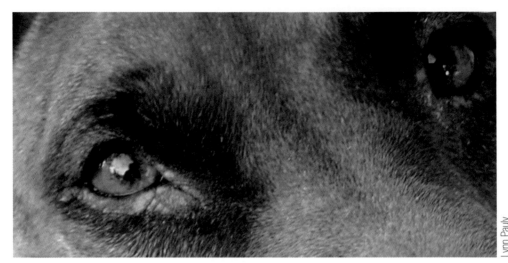

Lynn Pauly

Paula's Eyes

Another reason for pit bull spay-neuter legislation is that altered dogs are incomparably less aggressive. Because fatal pit bull attacks are almost always by unaltered animals, insisting on the spay-neuter of all but registered show pit bulls would keep the public safer, and thereby help kill the anti–pit bull hysteria.

For the same reason—aggression control—forced pit bull spay-neuter would help control the dogfighting world. Which brings us to one of the media's favorite topics of 2007, Michael Vick. In April 2007, raids of his property turned up evidence related to dogfighting, including sixty-six dogs (mainly pit bulls), treadmills to condition dogs, a "break stick" used to pry open fighting dogs' mouths, and a "rape stand" to hold female dogs for breeding.[12] Later searches of Vick's property revealed the carcasses of at least seventeen dogs buried in shallow graves.[13] He was accused of drowning and hanging dogs who did not appear to be good fighters, and supervising the wetting down and electrocution of a pit-girl after she lost a fight.[14] After hours of FBI grilling and failed polygraph tests, Vick admitted that he had killed dogs.[15] If only we could have persuaded the judge to dole out his just-under-two-year felony sentence in dog years.

Yet part of me says, "Heaven bless Michael Vick and his celebrity!" For years animal advocates who have known about dogfighting's ugly underbelly have been trying to get prosecutors and the public to take it seriously. We begged Nike to pull a 2003 ad tagged "The Battle," which included a shot of a pit bull and a rottweiler facing off.[16] Nike refused. Now Nike's message that dogfighting is cool and appropriate promotional material has come back to bite the company in the butt. Nike has pulled its Vick items—as if anybody were still buying them.[17]

cut off His balls?

No, I'm not asking about Michael Vick's! And I don't just mean pit bulls either. Animal advocates widely encourage the spaying and neutering of all pets. Of course there are ethical issues around surgically altering animals to control overpopulation. We would rather push safe sex, but if you think fingernails are rough on condoms, wait till you see what claws do.

There seems to be something sad about never allowing female animals to have babies. But perhaps those who choose to let their animals have litters have never thought how it must feel to have all of one's babies disappear right after nursing. Though we cannot know exactly how the animals feel, those of us who live with animals know some of the ways in which their emotional lives are similar to ours. We know we would be kidding ourselves to presume that they

don't feel loss when their babies are taken. So are we really being kind if we let them produce offspring that we will remove?

Some people cannot bring themselves to neuter their animals. Indeed, it would be good to see less invasive procedures become more widespread. Vasectomies, for example, would allow animals to engage in sex, a natural and enjoyable activity, without consequence. But leaving an animal entirely in his natural state is currently not an ethical choice; it is cruel to the unwanted offspring he might produce, or cruel to him if you do not plan to let him produce any. If you were never going to be allowed to have sex, wouldn't you prefer to be without the urge? Perhaps that accounts for all the married people on sex-chilling Prozac.

Doggie prozac?

When I first drafted this chapter, I wrote, as a half joke, that Prozac could be the perfect solution for pet overpopulation; Fido could come when you call but not any other time. And I suggested, not so jokingly, that it would also take the edge off all his lonely days in the backyard. Then I read in the *New York Times*, "Dogs with separation anxiety are now commonly treated with psycho-pharmaceuticals."[18] Drugs instead of company—goodness gracious. Dogs are pack animals. There is nothing wrong with your dog if he hates to be alone—go to the pound and get him a buddy.

Assuming, hopefully, that people don't widely start to substitute doggie Prozac for socialization, we best stick to spaying and neutering until less invasive procedures are common.

Females First

We are wrong to talk about spaying and neutering as if they are equally important. Imagine you have one fertile female and ten potent males. How many litters will you have a few months later? One. Imagine you have ten fertile females and one happy potent male. How many litters will you have a few months later? Ten. In fact, if you have a hundred fertile females and only one potent male, if he is up to the task you might end up with a hundred litters. (Just ask any musician who has spent a few months on the road without condoms.) As long as you have just one intact male, the number of litters depends entirely on the number of fertile females. So neutering a male, theoretically, has no impact on pet population in a neighborhood

where there are any unneutered males. Conversely, every single female who is spayed reduces the number of young of that species. That means that government and nonprofit efforts to reduce pet overpopulation should focus on spaying. And it means that the macho men, or perhaps animal rights ethicists, who won't neuter their male pets might not be having as much impact on pet overpopulation as we would have assumed.

To kill or not to kill?

People are sometimes shocked to learn that some organizations known for defending animal rights also kill animals. I feel I came to understand that position better when I learned that the head of one such organization races cars for sport. She risks her life for excitement, indicating that to her, life without stimulation would be worse than death. She believes in granting animals the right to the death in comforting arms she would surely choose for herself over a lonely and fruitless life in a cage. Moreover, she does not wish to shelter herself from the current harsh reality of pet overpopulation and let somebody else clean up the mess.

But is it not odd for advocates to kill those for whom they advocate? On one of the animal rights e-lists, the activist Adam Weissman proposed a compelling parallel, which I will paraphrase: Imagine a house for juvenile delinquents. Some of the kids are constantly in trouble with the law. Therapy has had little impact. They have been placed in foster care many times but it has never worked out. Some of the kids have even shown violence toward the families with whom they have been placed. The staff is therefore unenthusiastic about attempting more placements. Because suitable homes cannot be found, it appears that the kids are condemned to life in the high-security wing of the juvenile home, probably moving later to a similar institution for adults—a miserable existence.

Imagine then that the director of the home suggests that, rather than warehouse the kids in the hope of finding some ephemeral ray of light, his staff should face reality. He suggests that the best solution would be to humanely euthanize the teenagers. He would be fired on the spot. And rightly so.

Why then is such a solution considered appropriate when dealing with members of other species? In some cases, it is surely a compassionate choice made by those who would prefer that for themselves. But generally, that solution is considered acceptable only because in human society we do not take the lives and deaths of other animals as seriously as we take our own. Whether or not that is right or wrong is a complex ethical question beyond the scope of this chapter. But I think it is fair to write that those of us who have appointed ourselves as the defenders of other species should not turn around and treat them in a way that society would never treat humans. It is a sad truth that while we fight for practices and laws to end pet overpopulation, companion animals are being killed for lack of homes. Yet I think it is best for animal rights activists to release those animals for whom we cannot care, and, indeed, to leave the dirty work to somebody else. Otherwise, be it accurate or not, we give the impression that animal advocates are willing to kill our own constituents because even we hold animal life in little regard.

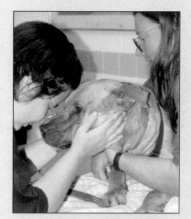

"They're good dogs. Which is exactly what the shelter workers assure them of as their lives are ended."

Photographs from the book *One at a Time: A Week in an American Animal Shelter,* © No Voice Unheard.
www.NoVoiceUnheard.org

While I do not blame shelter workers for feeling forced to do the heartbreaking job of killing unwanted animals, I would like to see what would happen if those with power to direct programs simply refused to keep killing. In the short term we would be sentencing animals to horrendous living conditions in overcrowded shelters. But if there were nowhere to put unwanted animals, the legislature would have to vote for the population control laws that animal advocates are continually proposing. As long as we keep killing, legislators can ignore the crisis and rely on those of us arguing for change to work against our own case by continuing to clean up and bury the evidence of dysfunction.

I do believe it is time for animal advocates to force change by refusing to kill companion animals, but I know this is a complicated issue and that a different stance can also be ethical. It saddens me when animal advocates who disagree accuse each other of not caring. Let's remember that we are all on the same side, all working toward the day when no innocent animal is killed for lack of a home.

The Bound Leading the Blind

Should there be guide dogs for the blind? Here's an issue where animal rights activists skirt the edge of misanthropy. If we are against the use of guide dogs, don't we care more about animals than people? If we are for them, aren't we supporting animal slavery?

A core problem is that as millions of dogs die for lack of homes, guide dog associations contribute to the overpopulation crisis by breeding more. The idea that only purebred Labradors are suited for the job is a myth, a myth smothering under the heap of Labradors bred only to be discarded because they didn't make the cut. Every also-ran is adopted out to a family, perhaps sentencing to death a pound dog who could have ended up in that family. Labradors are a dime a dozen at many pounds.

Many of the service dog groups have realized that one can find suitable animals among the shelter dog community. Young homeless dogs can be tested for the temperament and skills needed for the job. No dog has more incentive than a pound dog to

"Cashew, my fourteen-year-old yellow Lab, is blind and deaf. Her best friend is Libby, seven, her seeing-eye cat. Libby steers Cashew away from obstacles and leads her to her food. Every night she sleeps next to her. Without this cat, we know Cashew would be lost and very, very lonely indeed" (**Terence K. Burns**).

show up for the interview on time and with a good attitude! Testing pound dogs might be less convenient than breeding likely candidates and dumping the failures, but the ethical benefits should far outweigh any inconvenience.

Is it right to use the dog in that way, regardless of the source?

While spending time in New York I saw a blind man who makes his paltry living walking up and down the trains begging. Loud music blares from a boom box that hangs by his hip. Also hanging by, his head just a few inches from the awful din of the boom box, is the man's dog. He cowers with his tail so far up between his legs it is flat against his belly.

As they make their way up the car, my concern for the man's plight is matched by concern for the desperate animal. The dog is some kind of husky or collie mix, with a wolflike face and coat. It is easy to imagine him in the natural wild-canine state, running through fields, hunting and playing with his pack. While I feel for the man, I also feel that he has no right to do that to another animal.

The life of a guide dog, however, need be nothing like that. We have all seen blind people who adore their dogs and give them a wonderful life. The life of some guide dogs, who accompany their beloved masters everywhere (I choose the word "masters" carefully; guide dogs really are part of a benevolent slavery situation), are no doubt far better than the lives of millions of dogs who spend every day alone in the prison of the family backyard.

Many animal rights activists would say that even if the dog has a good life, we have no right to make a slave of him. But I find myself always getting back to the question, "Are we providing a life such that the animal is better off for having been born?" In the case of the miserable animal trudging up and down the subways with music blaring in his ears, I doubt it. In the case of the beloved guide dog who sits at his master's feet all day as she works at her desk, or sits at her feet in restaurants, or accompanies her on long walks and shopping trips, probably.

Much depends on the attitude of the human, which can often be summed up by the sign the dog is wearing. Does it say, "Don't pet me, I am working," in other words, "I am not a sentient being with my own emotional life—I am here for no reason other than to serve this human"? Or does it say, "Ask to pet me, I am working"? That would tell us the dog's first responsibility is to the person he is serving, but that warm interactions with others would be appreciated.

Another important question: Does the animal have any fun of his own in life? What a difference it could make if the person could sometimes meet a sighted friend at the dog park, and in the friend's safe care, could let the dog play with other dogs for a while—no doubt keeping a watchful eye on his human sitting safely nearby.

Like so many issues, this one is gray. Few animal rights activists would be incensed at the sight of an adopted animal, clearly loved, accompanying a blind person everywhere and being showered with affection. And those with the least interest in animal rights are unlikely to insist that the animals cannot be adopted rather than bred, or should be put in situations that are clearly pure misery. In the gray area of guide dog service we can surely find a place where compassionate people will find ways of rewarding the canine species for the service provided—we might call it "thanking the doggie"—so that the interaction can have true reciprocity.

Making Your Dog a Swan

The city of Rome made news around the world when it banned the clipping of dog tails and ears for cosmetic reasons, and the declawing of cats. Such ordinances are still few and far between—still almost unheard of in the United States.

Ear and tail cropping are physical mutilations for the sake of fashion.

Tail cropping removes an important communication device. It is a painful procedure, long accepted due to the now disproved myth that newborns do not feel pain. If the procedure is done by a veterinarian, the tail is clamped a short distance from the body, and the portion of the tail outside the clamp is cut away, with no anesthetic. Breeders may dock the tail themselves by tying it off, cutting off the blood supply. The dead portion falls off in a few days. The veterinarian Jean Hofve from the Animal Protection Institute writes, "This can be likened to slamming your finger in a car door—and leaving it there."[19]

Even the American Veterinary Medical Association (AVMA), a trade group for veterinarians that rarely takes the side of the animals against the financial interests of vets, has issued a statement recommending the discontinuation of tail docking for cosmetic purposes.

Ear cropping is a painful operation done on an area with an extensive blood and nerve supply. Like tail docking, it serves no purpose other than to conform to American Kennel Club standards.

Karen Dawn

kitty mob makeovers

Declawing is not like cutting your fingernails or even removing them completely. It is the equivalent of cutting off the tips of your fingers at the first joint. It is more accurately described as "partial digital amputation"—a torture technique sometimes employed on prisoners of war and zealously re-created in many mafia-themed movies. Who knew Hollywood scriptwriters were getting their ideas at the local veterinary office?

Dr. Hofve writes, "Cats waking from declaw surgery will thrash from wall to wall in the cage, howl, and shake their feet as if trying to fling them away."[20]

The procedure can cause chronic lameness, arthritis, and other long-term complications, including litter-box avoidance; the gravel hurts and the animal has trouble balancing.

If one wants cats, one probably has to make certain sacrifices—such as parting with leather furniture. And you know I'll be recommending that anyway later on in this book—in chapter 4's section on leather. One can now also cover cats' claws with soft plastic caps, and purchase scratching posts where cats can be encouraged to act out their natural behaviors.

West Hollywood was the first city in the United States to ban declawing. Of course, those in West Hollywood can still go to nearby Santa Monica to get their cats declawed—until Santa Monica brings in similarly compassionate legislation. Some activists use that kind of scenario to suggest that animal protection laws that cannot be enforced are useless. As people go out of their way, however, to get around an anticruelty law, it seems likely that they will be forced to confront whether or not they are doing the right thing by their animals. Those who previously would not have thought twice about having their cats declawed might reconsider if it is illegal to have it done in their city—even if it is still legal to get the procedure done elsewhere. Simply by being on the books, enforceable or not, an animal-friendly law can make people think about issues and change their behavior.

The Dog-Eat-Dog World of Dog Food

Mutts © Patrick McDonnell/King Features Syndicate

As I write this chapter, Paula, my vegetarian pit bull, is ten years old and in perfect health. Why vegetarian?

Many animal advocates cannot justify saving the lives of one or two animals at the expense of all the other animals their animals will eat, when it is easy to provide a delicious, nutritious, high-protein vegetarian diet.

As the idea of vegetarian dogs is relatively new and unusual, those of us who pursue it were pleased to read the following report in the UK's *Western Daily Standard*:

> A West vegan dog who was believed the oldest in the country died at the grand old age of 26, it was announced yesterday. Bramble the border collie clocked up more than 170 years in human terms after a lifetime diet of organic lentils, rice and soya.[21]

Incidentally, Bramble was a rescue dog.

The activist Jane Garrison, who headed up the Hurricane Katrina pet rescue effort in New Orleans, lost her own beloved vegan dog Codi, in 2007, at age seventeen.

One look at dogs' teeth tells us that they are naturally omnivorous, so I believe that while vegan diets seem to be fine, it is not necessarily the vegan diet that serves dogs so well. Dogs also do well on nonvegan wholesome foods. But animal rights activists have always been amused when people who feed their own animals commercial pet food express concern about dogs being vegetarian or vegan. That amusement turned to sympathy during the 2007 Menu pet food crisis, when Menu Foods recalled pet food being sold under more than one hundred brands after it was found to cause kidney failure. The company has confirmed sixteen animal deaths—animals who died in the company's own laboratory tests after Menu began receiving complaints. But the *Washington Post* tells us, "The FDA acknowledged last week that it's likely that more than sixteen dogs and cats died from eating bad food" and "meanwhile, a large veterinary hospital chain said it saw a 30 percent increase in kidney failure among cats during the three months that contaminated pet food was on the market, supporting the belief among pet owners and animal doctors that adulterated food has sickened or killed far more pets than officially recognized."[22] A *San Francisco Gate* article by a contributing editor for the Universal Press's nationally syndicated "Pet Connection" told us:

Codi Garrison (and carrot) 1990–2007

Jane and Mark Garrison

We at Pet Connection knew that the ten to fifteen deaths being reported by the media did not reflect an accurate count. We wanted to get an idea of the real scope of the problem, so we started a database for people to report their dead or sick pets. On March 21, two days after opening the database, we had more than six hundred reported cases and more than two hundred reported deaths. As of March 21 (2007), the number of deaths alone was at 2,797.[23]

Those of us feeding our dogs home-cooked veggie meals were deeply saddened by the grief of those who lost beloved companions, but were also relieved that we personally had nothing to fear.

Recall or no recall, commercial pet food can be awful stuff. It is generally full of slaughterhouse by-products—everything not considered fit for human consumption: the fat, gristle, cancerous tumors, and spinal material that holds mad cow disease. It may also include ground-up dogs and cats who have been killed at animal control facilities and sent to rendering plants, where their bodies are made into a protein paste that goes into pet food.[24] That's why animal rights activists feel comfortable feeding our dogs homemade or organic vegetarian cuisine.

Cats are carnivores and need an enzyme called taurine, which is available only in meat. If cats are to be made vegetarian, their diets must be supplemented with taurine. Many people keep cats on carefully supplemented vegan diets and the cats do well, but the issue is still debated within the animal rights movement. When I was once at a lecture with Peter Singer, he made the controversial and gutsy suggestion (as is his way) that vegans might prefer to adopt animals other than cats if they are not comfortable having meat in the house; he mentioned the many ex-laboratory rats who could be given homes. Rats are smart and affectionate pets. Though his stance was not popular among the cat lovers in the room, I think it worthy of consideration. If it is a cat's only option, surely a vegan life in a loving home is better than death. But one might consider giving that life to another deserving animal better suited for a strictly vegan household.

While dogs, who are naturally omnivores, not carnivores, can do particularly well on vegetarian diets, it is important to make sure the food you are giving them is nutritionally complete and balanced. A good resource of vegetarian health information for both dogs and cats is the Web site www.vegepets.com. Under "Dietary Information" there is a pull-down menu leading to various essays on vegetarian companion animal health issues and also a list of veggie pet food suppliers.

robin redbreast in a cage puts all heaven in a rage

The line above, from William Blake, really says everything we need to know about captive birds. For what crime do we condemn to cages animals whose natural right it is to soar across the heavens?

Up to forty million birds are currently estimated to be living as pets in American homes. Occasionally animal advocates have birds as pets because the birds need rescue. Because they can live eighty years, birds often outlive their human caretakers, or they are abandoned by them. In a front-page *Washington Post* story, Denise Kelly, a New York City parrot lover who founded the Avian Welfare Coalition to lobby for legislation protecting captive birds, comments:

Bizarro © Dan Piraro/King Features Syndicate

We love them because they're smart and they can fly and they're free. And then we stick them in a cage and make them completely dependent on us and clip their wings and take away their ability to fly. And then when they don't meet our expectations, we want to get rid of them.[25]

The *Post* article explains, "The instinctive behaviors that serve parrots well in their native habitats—screaming, biting, chewing and flinging food around—do not endear them to the typical American pet owner." It quotes the British parrot behavioralist Greg Glendell, who says, "Left isolated and under-stimulated, sensitive, intelligent beings tend to lose their minds."

Layla Love/lovephotography.org

"Children are born to roar with laughter as birds are born to soar across the sky"(photographer **Layla Love**).

Karen Dawn

The article includes a description of a room full of rescued parrots at the Foster Parrots sanctuary in Massachusetts. All day they yell at each other to "Shaddup" and "Knock it off." Those are the only words the parrots ever learned from owners who left them alone in cages all day, then couldn't be bothered with meeting their needs when they got home from work.

The Wild Parrots of Telegraph Hill is a multi-award-winning documentary about the relationship between a man, Mark Bittner, and a flock of wild parrots whom he tames and befriends without making captive. Rent it!

confining Nemo

It is questionable to write that fish are popular "pets," because they are rarely treated as pets at all. They seem to be acquired more as decorative objects. But they are not objects—they are animals who are far more intelligent than most of us have been led to believe. An article in *New Scientist* says of fish:

> In many areas, such as memory, their cognitive powers match or exceed those of "higher" vertebrates, including non-human primates. Best of all, given the central place memory plays in intelligence and social structures, fish not only recognize individuals but can also keep track of complex social relationships. . . . The structure of the fish brain is varied and rather different from ours, yet it functions in a very similar way.[26]

Does it not seem cruel to condemn an animal capable of keeping track of complex social relationships to life alone in a bowl? What about a tank? Many of us have been told that fish have memories of a few seconds, so every time they reach the end of their tank it is like they are arriving in a new place. *New Scientist*, however, quashes that myth by describing the following experiment:

> The fish were trained to locate a hole in a net as it approached down the length of a fish tank. After five attempts, they could reliably find the hole in the net. About 11 months later they were re-tested and their ability to escape was undiminished, even though they had not seen the apparatus during the intervening period.

"My word if it isn't old Witherspoon . . . Gosh it must be nearly . . . oooh . . . five . . . ten seconds . . ."

www.cartoonstock.com

The article also tells us that "larger groups learn to avoid the net faster than smaller groups because fish in a shoal learn by observing one another. Social learning opens the door for the transfer of information between individuals and from generation to generation."

When Jane Goodall discovered that chimpanzees use tools, a wall came down for many scientists who had believed that the ability to use tools separated us from other animals. Chimps were seen as much closer to us than had previously been thought. The *New Scientist* article tells us that some fish use tools for foraging, and some species of fish crush sea urchins with rocks.

The city of Rome recently banned the keeping of goldfish in barren goldfish bowls. Hopefully other places will follow suit. But even without legislation, perhaps, as we learn more about the intelligence of fish, people will start to question the ethics of confining fish to life in bowls or tanks, with little stimulation, when their natural environment is huge and rich.

Besides thinking about the misery of life in a tank for an intelligent creature, those interested in acquiring tropical fish should know about cyanide fishing. It is a common practice in the Philippines and Indonesia, which supply 85 percent of the tropical fish in the world's saltwater tanks. An article in *Scientific American*[27] explained that divers squirt cyanide at coral reef fish, disabling them so they can be caught before they hide in the coral. It is estimated that half of the fish die at the reef and then 40 percent of the captured survivors die before they reach the aquarium. The reef itself and nontarget fish are also killed.

Jurassic Bowl

An estimated eleven million reptiles live in U.S. households—that's not including lawyers. Some of the issues, both of acquisition and of their care, are similar to those for fish. Each year nearly two million live reptiles are imported into the United States. The trade kills animals and destroys habitats as collectors use such means as pouring gasoline into burrows to rouse the animals. A huge percentage of the reptiles die before they reach the pet store or dealer, often having been packed tightly together and shipped in

boxes. But as the animals came free to those who captured them, with no costs of rearing, feeding, or housing, high mortality rates do not interfere with the ability to make a profit. (It's kind of how some politicians look at soldiers.) Just one surviving animal in a box makes the shipment worthwhile. Further, it is estimated that 90 percent of wild-caught reptiles die in their first year of captivity because of physical trauma prior to purchase or because their owners cannot meet their complex dietary and habitat needs. What is life like for those who live? Animals who would generally spend their days hunting and foraging for food spend them instead sitting in tanks as curiosities for human spectators.

our interspecies families

A few in the animal rights movement might call for the phasing out of pets, the idea being that it would be better to let other animals live freely beside us, in harmony with us, than under our control in human society. But in the modern world, in cities, that isn't practical. And it is surely possible to keep pets in our homes and give them truly good lives such that they are better off for having been born. In order to ensure that, however, we will need for society to legally acknowledge something that most of us understand intuitively—that animals are not objects or property. They are other beings who become members of our family when we take them in.

In the aftermath of Hurricane Katrina, in 2005, thousands of people refused to evacuate.

Members of a frustrated rescue team simplified it for a *Dateline* news crew: They said people were refusing to be evacuated simply because "they won't leave their pets."[28] The rescuers were referring to those people still alive. Some people had died trying to weather the storm rather than going to Red Cross shelters, which refused pets. Public policy that refused to acknowledge nonhumans as family members killed people in New Orleans.

We saw clearly, during Katrina, how people feel about their animals. As houses flooded we saw whole families stranded on rooftops—whole families including the

animals. And we saw people wading through water, making their way out of the city, having left behind every piece of property but clutching their animals in their arms as they struggled to survive. Unfortunately we also saw shocking cruelty—elderly people who had lost absolutely everything being forced to leave behind their only companions as they boarded rescue boats and buses.

The one good thing that came out of that tragedy was that it edged us toward an understanding that we need changes in public policy. For starters, the government passed the Pets Evacuation and Transportation Standards Act, which forces states to provide for the evacuation of pets during emergencies in order to qualify for federal emergency funding. The act should safeguard against a recurrence of the Katrina situation and move us a step down the road toward official acknowledgment that animals are not mere property.

We see other steps in that direction in the field of animal law. Traditionally any harm done to one's animal could only bring compensation commensurate with the animal's monetary worth or cost of replacement. In the last few years, however, we have seen some cases in which people have been awarded tens of thousands of dollars for pain and suffering upon the loss of animals. The courts acknowledged that the animals' worth had nothing to do with their monetary value.

The Menu pet food crisis might have played out differently if animals had significant legal status. On the *San Francisco Gate* site, Pet Connection editor Christie Keith has presented a disturbing time line:

> Although there have been some media reports that Menu Foods started getting complaints as early as December 2006, FDA records state the company received their first report of a food-related pet death on February 20.
>
> One week later, on February 27, Menu started testing the suspect foods. Three days later, on March 3, the first cat in the trial died of acute kidney failure. Three days after that, Menu switched wheat gluten suppliers, and ten days later, on March 16, recalled the ninety-one products that contained gluten from their previous source.
>
> Nearly one month passed from the date Menu got its first report of a death to the date it issued the recall. During that time, no veterinarians were warned to be on the lookout for unusual numbers of kidney failure in their patients. No pet owners were warned to watch their pets for its symptoms. And thousands and thousands of pet owners kept buying those foods and giving them to their dogs and cats.

> At that point, Menu had seen a 35 percent death rate in their test-lab cats, with another 45 percent suffering kidney damage.[29]

During that time, animals were taken to vets with kidney failure, then taken home and given the same food. And though the recall wasn't issued till March 16, the *Washington Post* does tell us that the chief financial officer of Menu Foods sold about half of his stake in the company on February 26 and 27—three weeks before the widespread pet food recall.[30]

In chapter 5 we will see that culpable companies also perform poorly when human food is tainted, but the tardiness of the pet food recall was in a class of its own.

Now imagine the recall speed if Menu Foods thought it might face a multimillion-dollar lawsuit for every pet death. The paltry pace of the response we saw suggests instead a reliance on knowledge that pets are considered property. It appeared to rely on the inequity, or even iniquity, of law: that those who love animals can be reimbursed only for their market value, or at most for repair bills—the term we might as well give our veterinary costs while our animals have the legal status of objects.

A change in the status of pets would give them, and those of us who love them, some protection.

One of the most common reasons people cite for relinquishing animals to shelters is that they have been forced to move—often because of a change of job or a change in financial circumstance—and they have not been able to find pet-friendly housing. If society acknowledges pets as beings rather than as objects, and as members of our family, then excluding them from housing could bring discrimination suits.

While granting rights, a change in status should also bestow responsibilities upon those of us who have nonhuman family members. If animals were not legal property, they could not be bought and sold but would have to be adopted. As with the adoption of children, prospective adopters would be expected to meet requirements. Paritney wouldn't have a hope until she'd spent at least a year in rehab, and we would all have to demonstrate the ability to care for the animals over the full length of a lifetime, providing the animals not just with food and shelter but also with companionship and stimulation. Just as child services would prosecute parents who locked up a child alone all day every day, if our world had a truly evolved view of animals, animal services would prosecute similarly for that abuse. If one were unable to continue to care for an adopted animal, one could choose to give up that animal to an adoption agency, as one can with a child, but abandonment would be illegal. And anybody who had given up an animal without very good cause would not be allowed to adopt again.

Only a change in legal status for animals will allow for such developments. Animal rights activists would like to see that change for all animals. An obvious place to start is with our beloved pets.

The Washington Post, Saturday, September 10, 2005

"Best Friends Need Shelter, Too"
Op-Ed by Karen Dawn

The week after Hurricane Katrina hit, the media covered the thousands of low-income people trapped for lack of means to get out. Almost two weeks later, thousands still hadn't left, in many cases because official policy would not accept the bond between people and their nonhuman family members. Members of a frustrated rescue team simplified it for a *Dateline* news crew. They said people were refusing to be evacuated simply because "they won't leave their pets."

There is a class issue involved here. While Marriott hotels welcomed the pets of Katrina evacuees as "part of the family," people who had to rely on the Red Cross for shelter were forced to abandon that part of the family or attempt to ride out the storm. It cannot be denied that many poor people are dead as a result of "no pets" policies.

The *Los Angeles Times* reported on Patricia Peony, who wondered whether her son Billy had survived. She had begged him to leave, but he was afraid to abandon his animals. CNN showed the rescue of a family, including a dog, sitting on a rooftop as a boat pulled up. The boat left without the dog. Staying with a dog and risking their own lives is not

an option for people who have children to provide for. The parents were given no choice but to abandon the dog, and to break their children's hearts. As they pulled away they all watched their trusting, confused and terrified canine family member alone on the roof.

At Red Cross shelters there are families that have lost their homes and all of their possessions but are thanking God that they are all safe. Others are frantic, unable to think of anything besides the slow death of beloved companion animals they were forced to leave on rooftops or at bus boarding points. One woman, with no other possessions left, offered her rescuer the wedding ring off her finger to save her dog, to no avail.

A young boy carried a dog in his arms as he tried to board a bus to the Houston Astrodome. Dogs were not allowed. The Associated Press story reported that "a police officer took one from a little boy, who cried until he vomited. 'Snowball, Snowball,' he cried." In a similar story, an old woman, traveling alone except for the poodle in her arms, was forced to leave him behind to wander the streets. We have read other stories of elderly people forced to choose between their lifesaving medications or their life-affirming pets. CNN's Anderson Cooper even reported on a woman, legally blind, who for 10 days had been told that she could not take her service dog with her if she was evacuated. She had stayed put until the CNN cameras arrived and the police relented.

Many large hotel chains, aware of the human-animal bond, now allow guests of varied species. Sadly, those organizations on which we rely, not when on vacation but in life-or-death circumstances, are not up with the times.

The pets pulled from people's arms would not have taken seats meant for humans. There is no reasonable explanation for abandoning them. They were the last vestiges of sweetness, in some cases the only living family, of those who had nothing left. But the police officers were just following orders—orders that reflect an official policy inconsistent with how people feel about their animals.

Red Cross shelters that do not have animal-friendly areas, or do not coordinate with humane groups to make sure that there are animal shelters nearby, are out of touch with the needs of a society in which 60 percent of families have pets and many view them as intrinsic members of the family.

"Pete Wentz and Hemingway" Courtesy of Pete Wentz

Fall Out Boy's **Pete Wentz** takes Hemingway everywhere he can. Here they are on tour together.

Robert Whitaker/Getty Images

chapter three

all the world's a cage: animal entertainment

A bloody history continues

In his section on elephants in *Natural History*, Pliny the Elder gives us a glimpse of the history of animals in human entertainment. He describes tightrope-walking elephants, including one who "was often beaten with a lash and was discovered at night practicing what he had to do."[1] Pliny also portrays, in distressing detail, a circus performance during Pompey's second consulship:

> Twenty elephants fought in the circus against Gaetulians armed with throwing spears. One elephant put up a fantastic fight and, although its feet were badly wounded, crawled on its knees against the attacking bands. It snatched away their shields and hurled them into the air . . . a second elephant . . . was killed by a single blow: a javelin struck under its eye and penetrated the vital parts of its head. All the elephants, en masse, tried to break out through the iron railings that enclosed them. . . . But when Pom-

pey's elephants had given up hope of escape, they played on the sympathy of the crowd, entreating them with indescribable gestures. They moaned, as if wailing, and caused the spectators such distress that . . . they rose in a body in tears. . . .

Reading that passage, our instinct is to be thankful that it is ancient history and that civilization has evolved. Yet in Spanish stadiums packed with tourists from around the world, we still see animal slaughter presented as art. In fact, many of the forms of entertainment widely considered to be good clean family fun rely on well-documented abuse and even bloodshed that takes place mostly behind the scenes.

The ABCDs of Animal Entertainment

In the various fields of animal entertainment explored below, we see similar problems. I call them the ABCDs of the animal entertainment industry:

Acquisition

Animals do not give themselves up willingly for our entertainment. They must be ripped away from their families.

Brutality

Unlike domesticated animals who would respond to positive reinforcement and affection, wild animals trained for entertainment are dominated and intimidated into submitting to human will.

Confinement

While their natural environment might offer them whole jungles or oceans to roam, captive animals are confined forever in cages, in tanks, in pens, or with chains, for our viewing pleasure.

Disposal

When they have outlived their economic usefulness, animals used for entertainment are disposed of in shocking ways.

behind the big top

The Meek Shall Inherit Nothing.

Anthony Freda

From Pliny's description of an ancient elephant circus, we turn to look at modern circuses, where elephants are one of the biggest attractions—and one of the most controversial.

In the United States, Ringling Bros. advertises on its Web site that none of its elephants "are taken" from the wild. They aren't currently being taken because a ban now forbids the capture of wild elephants. But according to the *Asian Elephant North American Regional Studbook*, last updated in 2005, the majority of Ringling Bros. elephants alive today were wild-caught.[2]

What does that tell us about the elephants' histories? Neither a mother elephant, nor any of the protective herd, will willingly part with a baby, so the best way to acquire

elephant calves is to slaughter an entire herd. African and Asian govenments sometimes talk about the need to cull their herds, but that need is driven by Western
purchase of the babies; it makes the culling lucrative. In the year 2000, hunters in
Africa killed sixty-three elephants so that their babies could be collected and sold to
the entertainment industry.

Early life is not much easier for circus elephants born in captivity in the West.
Whereas wild elephants spend their whole lives with their mothers, grandmothers,
aunts, and cousins in matriarchal herds, Ringling Bros. and other circuses separate
the young elephants and start their circus training at age two. United States Department of Agriculture (USDA) inspectors have found rope burns on the babies' legs,
inflicted as they were forcibly pulled from their mothers.[3]

An Elephant Never Forgets a Bullhook

Training a wild animal involves brutality. The circus public-relations departments tout the use of "positive reinforcement" as if they were training domestic dogs. Dogs have evolved over thousands of years to live with humans. They submit in return for treats or affection. If that were so for wild animals, we would see trainers in circus rings with bags of treats. Instead we see whips and bullhooks. A bullhook is a metal club with a sharp hook at the end designed to dig into an elephant's flesh.

Undercover video shot behind the scenes at the Carson & Barnes Circus shows an elephant trainer with his bullhook-wielding protégé. The trainer yells: "Tear that foot off. Sink it in the foot. Don't touch them, hurt them! . . . Make them scream. . . . If you are scared to hurt them, don't come in the barn."[4]

As he yells, "Sink that hook in," he shows his trainee how to wiggle it around in the flesh and says, "When you hear that screaming, then you know you've got their attention. You're the boss."

He explains, "I am not going to touch her in front of a thousand people. She's going to do what I want."

True—because an elephant never forgets how that bullhook felt in that training session.

That video can be viewed online at www.Circuses.com. We also see young elephants jabbed with bullhooks and electric prods as they practice the familiar tricks, like running in a circle while holding each other's tails, which we, as innocent and unsuspecting children, so loved to watch.

The Carson & Barnes scenario is not unique. Undercover video shot at Ringling Bros. also shows trainers whacking elephants with bullhooks.[5] Four animal protection groups, the Fund for Animals (FFA), the American Society for the Prevention of Cruelty to Animals (ASPCA), the Animal Protection Institute (API), and the Animal Welfare Institute (AWI), with undercover footage and eyewitness testimony from the former Ringling Bros. employee Tom Rider, have brought a lawsuit under the Endangered Species Act (ESA) against Ringling Bros. for its mistreatment of Asian elephants. (You can read some of the details of that suit at www.awionline.org/wildlife/elephants/rbsuit.htm and also view some shocking footage.)

We know something about the trauma experienced by wild-caught elephants. The information on the Animal Welfare Institute Web site about Ringling's treatment of its captive-bred baby elephants is equally disturbing. We learn that the suit against Ringling refers to what federal inspectors termed "large visible lesions" on the legs of two baby elephants, Doc and Angelica, and notes that Ringling has explained that the lesions were inflicted during the "routine process of separation" of baby elephants from their mothers. The babies are pulled from their mothers with ropes and chains tied around their necks and legs.

KTVU in San Francisco ran a report on the suit, which you can view online at www.ktvu.com/video/4936923/detail.html. It is well worth watching. It includes video received during the discovery period of the federal lawsuit. We see the death of the baby elephant Benjamin, while the reporter's voice-over tells us, "The lawsuit claims that Benjamin drowned when he tried to avoid a bullhook that had frequently been used on him by trainer Pat Harned. It is Pat Harned you see hitting him in the water." The video shows the trainer striking Benjamin with the bullhook while he is in the water. Then the announcer says, "These are his first words when he realizes Benjamin is dead," and we hear Harned yell, "Oh God," as we see him prod at Benjamin's lifeless body.

That KTVU story also shows footage of a young elephant, Shirley, giving birth while chained by three legs, tugging and pulling at the chains, "standing in her own amniotic fluid on a concrete floor," and unable to move around at all as she naturally would during the process, or even to kneel. The baby therefore drops many feet onto the cement floor. We learn that the baby, Ricardo, at eight months of age, breaks two legs in a fall and is euthanized. We are told, "The animal rights groups contend it occurred while he was being trained to sit on a stool for Ringling Bros. performances." The reporter also tells us that Ringling Bros. has contended that the little elephant had a bone disorder and broke both his legs during a play-yard fall. Unfortunately there is no video of that fatal mishap.

The KTVU report continues, "Ricardo is one of three baby elephants who recently died under controversial circumstances under Ringling's care."[6]

The KTVU story told us about Benjamin and Ricardo—the other was Kenny. An *Orlando Sentinel* front-page story opened with: "Kenny the elephant, too weak to stand or eat, died in Jacksonville, bleeding from an infection after performing his third Ringling Bros. and Barnum & Bailey show of the day, according to a report released Wednesday by three animal-protection groups."[7] The *Washington Post* has quoted the USDA on Kenny's case, telling us: "In a 1998 case, Ringling Bros. made a sick baby elephant named Kenny perform 'after determining that the elephant was ill and needed to be examined by a veterinarian,' according to the USDA. After the show, Kenny died. Ringling Bros. paid a civil penalty of $20,000."[8]

Chains, Trains, and Automobiles

When not being bullied in the ring, and forced to watch unfunny clowns that scare children, circus animals are lugged from town to town in tiny trailers. Or they are confined, between performances, in cages or on chains, or in small pens behind electric fences. A circus elephant spends much of her life chained by one leg. In 1993, Ringling Bros. helped defeat legislation in California that would have limited the number of hours per day that elephants could be chained.

small bear at roadside zoo

© 2007 by Sue Coe

After their youthful performing days are over, circus animals are put on display at zoos. For an elephant, that means being taken from the herd with which she has worked for years, the only family she has known since she was a baby. These intelligent and highly social animals often spend their remaining years standing alone in concrete pens.

Karen Dawn

zoos—It's No Jungle in There

Captivity itself, life in cages, is the most obvious problem for animals in zoos—no small issue, as human society locks up its own members only when they have committed crimes. Zoos are improving, and we have all seen or heard about large natural enclosures at some of the world's best zoos. Yet even when visiting a zoo with the terrific reputation of the famous San Diego Zoo, one finds pairs of monkeys who spend every minute of their lives in cages no bigger than maximum-security jail cells so that visitors can get a close look. If only people would settle for a video and a package of fresh monkey poop. Actually, it isn't really fair to compare the monkey cages to jail cells—at least murderers and rapists get an hour a day in the exercise yard.

For the animals who require the most space—elephants—zoo captivity is becoming the most controversial. In the wild, elephants walk up to fifty miles per day. They come from warm climates and fare particularly poorly through cold winters, which they spend shut up in barns, standing all day on concrete floors. Standing on concrete destroys their feet. When standing becomes too painful and an elephant collapses, her weight crushes her internal organs; so foot ailments eventually prove fatal.

SIPRESS

"He didn't do anything, Gregory. This is a zoo."

Ron Kagan, the director of the Detroit Zoo, made worthy headlines after he decided that his zoo did not have the capacity to care properly for elephants. He announced that he would release his zoo's elephants, who had started to suffer from foot ailments, to the PAWS sanctuary. Wanda and Winky are now flourishing on hundreds of acres in Northern California. As the controversy around elephants in zoos has escalated, a few other zoos have followed Detroit's lead, but hundreds of elephants in the United States still languish in pens that cannot meet their needs.

The Elephant in the Room

Besides acquiring elephants from circuses, zoos, like circuses, may import them from Africa and Asia. Take Maggie, for example. An *Anchorage Press* cover story told us how Maggie had come to America: "In 1983, a Zimbabwe cull left five baby elephants watching on grassy plains as all the adults in their herds, all the elephants they'd ever known, were cut down around their ears. The five orphans were purchased by Americans and flown to the Catskill Game Farm, a private zoo in Upstate New York."[9] One of them, Maggie, was then flown to Alaska as a companion for the Alaska Zoo's solo elephant, who has since died.

Maggie lived at the Alaska Zoo in Anchorage for more than twenty years. She stood alone in a barn through the long cold winters, in poor health and poor spirits, in need of companionship and exercise. While activists campaigned for her release to an elephant sanctuary where she would roam hundreds of acres with other retired elephants, the Alaska Zoo's attempted solution was to provide her with a specially made elephant treadmill. The press coverage focused mostly on the funny spectacle of an elephant on a treadmill, rather than on the tragedy of an elephant destined to live and die alone in a barn.[10]

In 2007, twice within a few days, Maggie lay down and was unable to get up without the help of the local fire department.[11] The media stopped chuckling. In fact, the ensuing coverage brought an international outcry and the zoo board finally agreed to release Maggie. They dillydallied for months over when and to where they might let her go—procrastinating even after California's PAWS sanctuary offered to pay for Maggie's airfare, retired TV show host Bob Barker offered to kick in $750,000 toward her life care, and an anonymous donor promised the zoo $100,000 if Maggie were released to PAWS.[12] Finally, in late 2007, PAWS prevailed. Now Maggie pond-bathes and sun-bathes at the sanctuary. You can watch her, live, at PawsWeb.org—just don't read too much into a book called *Thanking the Monkey* sending you to check out some gorgeous gal's webcam.

Spare the Rod and Spoil the Star

Though we associate animal training with circuses, zoo animals do not necessarily escape it. Some zoos function as mini circuses, putting on shows to attract tourists. Wild animals may be brutally trained just to be made manageable. A *Los Angeles Times* article tells us, "For a hundred years, the accepted way to manage elephants in zoos was through close contact and dominance, including whacking the mammoth mammals with sticks or ax handles when they were balky or cantankerous."[13]

It continues, "Conventional wisdom, derived from elephant handlers in Africa and Asia, held that to spare the rod was to endanger the keeper and that, as wild beasts, elephants need to be intimidated into submission."

Thai Torture

In Thailand, that intimidation comes in the form of a traditional practice called phaa-jaan. A baby elephant is held in a body-fitting pen for three days. She is stabbed with sticks in her most sensitive spots, such as inside her ears, and she is whipped and beaten. There is a good chance—about 30 percent—that she will die during the breaking ritual. The elephant's mother is chained nearby so that the baby learns that when she screams, her mother will not come to help, and that she is utterly at the mercy of humans.

You can see footage of phaajaan at www.HelpThaiElephants.com. If you have ever been tempted to ride an elephant on vacation, or support an industry that imports Thai elephants, it may give you pause.

Out with the Old

Animals owned by zoos are treated as objects to be bought, sold, and loaned, with no regard for social bonds. Young active animals are more popular with zoo visitors than older animals—so older animals are dumped.

Rarely does the issue get as much publicity as it did with Peaches, Tatima, and Wankie, who were once attractions at the Zoological Society of San Diego's Wild Animal Park. In 2003, to make way for seven young wild elephants caught in Swaziland, the renowned San Diego Zoo shipped the three older elephants off to Chicago's Lincoln Park Zoo. Animal rights activists had urged the zoo to instead relinquish the elephants to the PAWS sanctuary in Northern California. They warned that moving the elephants to a facility where they would stand in barns for months on end through icy winters would be a death sentence. The activists were accused of hysterical hyperbole. Within two years, Peaches, Tatima, and Wankie were dead.

In an article published in *U.S. News & World Report*, Michael Satchell writes, "Dumping animals is the big, respectable zoos' dirty little secret."[14] He describes two gibbon monkeys, who "were discovered by a reporter one recent broiling day in a filthy cage with no water and a few scraps of rotten fruit" at a Texas roadside zoo. The gibbons were castoffs from one of the nation's top wildlife institutions, the Rosamond Gifford Zoo in Syracuse, New York.

Satchell writes: "Their plight points to a little-known practice by some of the nation's premier zoos: dumping surplus, old, or infirm animals into a vast, poorly regulated—and often highly profitable—network of substandard, 'roadside' zoos and wildlife dealers who supply hunting ranches and the exotic-pet trade."

Canned Hunting

The hunting ranches to which Satchell refers host "canned hunts." Hunters pay set fees for guaranteed kills of exotic trophy animals, many of whom are the older castoffs of the entertainment and exotic-pet industries.

Can the Canned Hunts

Wanted—Dead or Alive

Occasionally, zoo animals are shot without being "retired" to hunting ranches. A story on the front page of the *Dallas Morning News* told us that when teenagers taunted and threw stones at Jabari, a gorilla at the Dallas Zoo, he jumped the wall of his enclosure. As he fled, he tackled people in his way. Despite having the strength of many men, he inflicted only minor injuries—yet he was shot dead. The teenagers suffered no consequences.[15]

Curiosity Killed the Chimp

The *Lincoln Journal Star* reported that three chimpanzees were shot and killed by the director of Zoo Nebraska after they escaped from their cage. No people had been hurt during the escape.[16]

When a chimpanzee was shot dead after escaping from the Yorkshire Zoo, the *Yorkshire Post* reported, "The spokesman said the animal was shot on zoo premises and Flamingo Land bosses stressed that no member of the public had been in any danger."[17]

After the chimps Coco and Jonnie tunneled their way out of their enclosures at the Whipsnade Zoo in Bedfordshire, UK, Coco was easily rounded up but Jonnie, according to newspaper reports, "was more determined." So he was shot dead. The zoo spokesperson explained, "That is normal practice if a chimp cannot be recaptured" even though "at no stage was the safety of our visitors at risk." In fact, newspaper reports tell us that "all members of the public were ushered out of the zoo before the keepers attempted to round up the chimps." It is hard to understand why, with the public safely out of the zoo, it was deemed necessary to shoot a chimp who could not be captured quickly. But then perhaps the zoo was shaken by its recent history of human injury: A few months earlier a worker was hospitalized after falling into the lions' enclosure. The lions were locked in their dens, but the worker suffered a cardiac arrest.[18]

Please Don't Eat the Animals

Perhaps the most shocking wake-up call for those who see zoos as places that love and protect animals came when Thailand's Chiang Mai Night Safari Zoo announced plans to celebrate its grand opening. It planned to hold an "Exotic Buffet" for VIP guests. The buffet would include tiger, lion, elephant, and giraffe meat. It took a huge public outcry to cancel that feast.

Urban Elephant

Occasionally there are happy endings for the animals used in circuses and zoos. Allison Argo's Emmy Award–winning PBS documentary, *The Urban Elephant*,[19] tells the story of Shirley, who arrived as a baby from Asia and spent thirty years in the circus till a leg injury ended her "career." Unlike a professional human athlete or performer, she didn't get a cushy job in a broadcasting booth. As thanks for her years spent entertaining children she was retired to the Louisiana Purchase Zoo. There she lived for twenty-two years, in a concrete stall with a small grass yard, never seeing another of her own kind. Female elephants desperately need the companionship of other elephants—but that was barely known when Shirley arrived at the zoo. In 2003, however, the people running the Louisiana Purchase Zoo decided they could no longer give Shirley a home and agreed to send her to what the documentary calls "a little piece of heaven carved out of the rolling hills of Tennessee." Carol Buckley's Elephant Sanctuary provides hundreds of acres on which retired elephants wander, free to come and go from the heated barn.

Solomon James was Shirley's keeper at the zoo for twenty-two years. He clearly loves her. In the film he says he spends as much time as he can with her so that she doesn't feel all alone, and we see him bathing her and talking to her with great affection. We watch as he accompanies her to her new home in Tennessee. Then with tears in his eyes he says to the camera,

> I am going to miss her. But when I saw this place I told her that there will be no more chains. She is free now. I don't know who was the first to put a chain on her but I am glad to know I was the last to take it off. She is free at last.

Then he says, "I am going to miss you, Shirley, my big girl."

The elephants at the sanctuary come into the barn one by one to greet the new arrival, who is at first in a separate pen. They touch trunks through the bars—it looks like kissing. When an elephant named Jenny arrives, there is an uproar. Carol says she has never heard anything like it. The trumpeting goes on through the night. When Carol comes into the barn the next morning she finds that in their desperation to get close to each other, Jenny and Shirley have bent the steel bars between them.

Carol did some research into their backgrounds and found that thirty years earlier they had worked together in the same circus. Jenny was just a baby, and Shirley would have been like a surrogate mother.

At the sanctuary they became inseparable. We see them walking, resting, and bathing together. The documentary's final shot is of them standing in the middle of a sunlit field, looking together at their new home, with their bodies pressed against each other. They have woven their trunks through each other's front legs as if they are attempting to bind themselves together.

Their tale has a bittersweet ending. A few years after Shirley's arrival, Jenny died young. Thank heavens she lived long enough for the two to reunite and spend peaceful time together. On the Elephant Sanctuary Web site you can read of Jenny's passing and of Shirley and others standing by till the end. Shirley has become the matriarch of the Elephant Sanctuary herd.

The final words of Carol Buckley's tribute to Jenny can only make us long for the day when all animals live with the love meant for all beings:

> In honor of Jenny we will play to our hearts' content, sing unabashed to the open skies and live each day with a joy that Jenny taught us. We will not focus on sorrow, only thankfulness for the gift of having known Jenny.[20]

Shirley and Jenny bathing together at the Elephant Sanctuary

marine parks

Shame Shame Shamu

Marine parks are zoos or circuses in the water. When you visit them and pay twenty dollars for your penguin hat, you are supporting the same ABCDs of animal entertainment. The reporter Sally Kestin opened a weeklong front-page series for the South Florida *Sun-Sentinel* with the following description, which sums up "Acquisition":

> Four decades ago, hunters off the coast of Washington found the perfect young killer whale specimen swimming with its mother. They fired a harpoon, hoping to attach a buoy to the bigger animal that would make trailing them easier. But the spear went in deep and the mother whale drowned. The crew made a deal for the young whale with SeaWorld. The company today says it did not know about the capture, but it did calculate correctly that crowds would come to its San Diego park for the chance to see a killer whale up close.
>
> The modern marine park industry began with the killing of Shamu's mother.[21]

The dolphin industry also combines capture with killing. Every year there is a massive dolphin slaughter in Japan. ABC's *Primetime* covered the slaughter and showed us, according to correspondent Chris Cuomo's voice-over, "dolphin trainers separating out the best of the catch for transfer to parks around the world," while the rest of the dolphins were "ruthlessly slaughtered by fishermen for their meat."[22]

Captured live dolphins can sell for up to $100,000, whereas dead dolphins sell for a few hundred dollars. So live capture is a vital part of the dolphin slaughter industry. A *Japan Times* article on the Taiji dolphin slaughter interviewed Ric O'Barry, who had captured and trained dolphins for the television show *Flipper*. He now campaigns against marine mammal captivity. The article includes the following:

> Every year, an unknown number of healthy young specimens are selected and removed from the killing coves to be sold into the inter-

national dolphin captivity industry, to be kept in aquariums, trained to perform at dolphinariums or for swim-with-dolphin programs. At Taiji, those involved appear to reap rich rewards in this way, and O'Barry said he was told there that the fishery drives would stop and those carrying them out would go back to catching lobsters and crabs if they were not offered huge sums for "show" dolphins.[23]

Theme parks outside the United States, where Westerners vacation, actively support the dolphin capture industry. The United States and Canada currently have bans on wild-caught dolphins, but the industry finds loopholes. For example, Vancouver does not allow the importation of wild-caught dolphins, but in 2005 the Vancouver Aquarium was able to import two wild-caught dolphins from an aquarium in Japan after the dolphins were deemed too injured to be rehabilitated for life in the wild. The Vancouver Aquarium spent about $200,000 on the purchase and transport. The Japanese aquarium is not bound by laws against dolphin capture, so the dolphins sold to Vancouver could be immediately replaced with healthy wild-caught dolphins. Therefore people paying the price of admission at the Vancouver Aquarium, or at any other aquariums or theme parks that have made similar deals, are subsidizing the dolphin capture/slaughter industry.

Your brain is even bigger than ours.

Perhaps that's why we don't find imprisoning you & forcing you to do tricks "entertaining."

Dist. by King Features

BIZARRO.COM

Avik Gilboa/Getty Images

After surfer-activists, including *Heroes* star **Hayden Panettiere**, attempted to interfere with the hunt, Hayden sobbed as she said,

"A baby stuck his head out and kind of looked at us, and the thought that the baby is no longer with us is very difficult."*

Go to *SaveJapanDolphins.com* to learn what you can do to help.

Photographs of Hayden Panettiere on the beach by Richard O'Barry. Courtesy of Richard O'Barry and Hayden Panettiere

Showbiz Tonight, CNN, December 14, 2007.

Courtesy of Sea Sheperd

The brutal training at marine parks does not involve electric prods or bullhooks. The parks tell us that they train their animals with "positive reinforcement," such as food rewards. In other words, they use starvation. Ric O'Barry worked at many marine parks. He explains in his book, *Behind the Dolphin Smile*: "If they are not hungry, there is no way to teach them anything. The hungrier they are, the better they learn."[24]

Once trained, dolphins must perform daily for their meals. That is why theme parks, and also television shows like *Flipper*, have to use more than one dolphin. O'Barry explains, "When they are full, you can't get them to do anything." When we watch marine mammals at theme parks doing the amusing tricks that we like to imagine they enjoy, they are working just to keep themselves alive in barren concrete tanks.

O'Barry believes captivity is harder on dolphins than on almost any other animals because they naturally live in complex underwater worlds and swim hundreds of miles per day. Those whose idea of exercise is the walk from their car to the Starbucks counter probably can't imagine the torment of that kind of confinement. But next time you down a venti soy latte and then get stuck in traffic for an hour, you might get just an inkling—then try multiplying the feeling by twenty-four, then by 365.

Lolita—Slave to Entertainment

Dumping is less of an issue for marine mammals; unfortunately for them, they don't lose their appeal to humans until their deaths.

A case in point is the Miami Seaquarium's solo orca. There is a superb, award-winning documentary on the fight to free her, called *Lolita: Slave to Entertainment*.

You can find out about the film, and watch a promotional segment, at www.SlaveToEntertainment.com.

The film follows activists who have been campaigning for years to have Lolita released back into Puget Sound, from where she, along with six other young family members, was taken in 1970.

During that capture, the whales were driven into a cove, and the young were separated from the others with nets run across the water by men in speedboats. People who live near the cove say that from miles away you could hear the awful sound of the captured whales screaming. Four baby whales drowned as they charged the nets to get to their mothers. The babies' bodies were carted out to sea, slit open and stuffed with rocks, roped to anchors, and sunk to the bottom of the ocean. Seven other young whales, including Lolita, were shipped to marine theme parks around the world. Though orcas will live to between fifty and ninety years in the wild, they generally live only to their teens in amusement park tanks. Lolita is the only one of the seven still alive. She is kept in a tank even smaller than what is recommended by the paltry regulations of the 1972 Marine Mammal Protection Act.

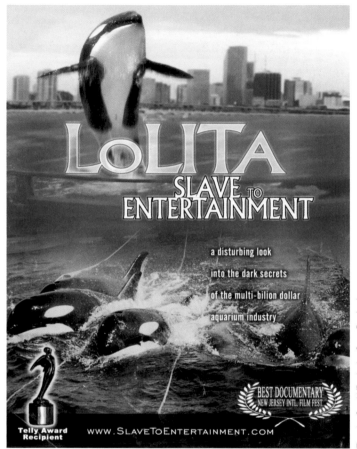

Timothy Gorski/Rattle the Cage Productions

Family members whom Lolita knew until the age of six, and in whose dialect she vocalizes, still swim in Puget Sound. Efforts to have her released back there, before she dies alone in a tank, have even included a benefactor's offer of $1 million to buy her freedom. But Lolita has already made $160 million for the Miami Seaquarium and is thus far too popular to be allowed freedom and family.

Hollywood stories

Tarzan and Jane Goodall

Fans of *Extras* can laugh at the way Ricky Gervais and his sorry cast of characters are treated compared to the A-list stars. But at least they aren't kept in cages and retired to vivisection laboratories. They have it a lot better than their nonhuman costars—particularly the chimps. In Hollywood the ABCDs of the entertainment business are perhaps seen most vividly with chimpanzees.

On the Chimpanzee Collaboratory Web site you can view a ten-minute film called *Serving a Life Sentence*, about the use of chimps in entertainment.[25] It features the

"*O.K., so no animals were harmed, but were they adequately compensated?*"

leading primatologist Jane Goodall, the primate behavioral researcher Dr. Roger Fouts, and the prominent film scriptwriter and director Callie Khori (*Thelma and Louise* and *Divine Secrets of the Ya-Ya Sisterhood*), who refuses to use chimp actors in her movies.

Jane Goodall talks about the training of chimps for the movie business. She fears that when people see that there has been a humane officer on the set they will assume that the animals were well treated. She explains, however, that most of the abuse happens before the chimps get to the set. She says, "Before that, most of the trainers want to establish a relationship based on fear so that they get instant obedience." She says that one method of behind-the-scenes training involves an iron bar surrounded by newspaper, then, "on the set you just need a rolled up newspaper."

The short film includes coverage of the famous animal trainer "Jungle Josh"

Weinstein, taken by KARE 11 Television in Minnesota. Even knowing the cameras were rolling, Weinstein threatens his chimp, Tarzan, with fury in his voice. You will find it particularly sad to watch the film knowing that Tarzan died under suspicious circumstances weeks after the story was shot.

It's Hard Out Here for a Chimp

When not on the set, animals used in the film business live in cages.

Mature chimpanzees, no longer cute, have the strength of many men and cannot be safely used by the industry. Though other Hollywood animals may end up on hunting ranches, it is illegal to hunt primates in the United States, so chimps are discarded elsewhere. They are often sold to roadside zoos as described above. Jane Goodall explains that respectable zoos don't want them because performers cannot fit into chimp troops. But bad roadside zoos don't have troops; they often house chimps in isolation. In the Chimpanzee Collaboratory film we learn about Chubbs, who now spends his days in a cage at a roadside zoo. He starred in Tim Burton's 2001 production of *Planet of the Apes.*

A few chimps may end up in some of the new sanctuaries being founded (described later in chapter 6), which, like the elephant sanctuaries, offer them some sort of decent life. Too often they end up in cages, like at Animal Haven Ranch, a sanctuary where a tragic event brought media attention to the plight of Hollywood chimps:

In February 2005, St. James Davis was at the Animal Haven Ranch visiting Moe, a chimp he had raised from a baby but had been forced to give up after the chimp bit off a neighbor's finger. During Davis's visit, two other chimps, discarded Hollywood movie stars, escaped from their enclosure and attacked the man, biting off most of his face, his testicles, and one of his feet. Nobody knows what brought on the attack. Perhaps the sad, bored, discarded animals were jealous of the attention given to Moe. Perhaps the rage that built up over years of brutal training for the entertainment industry was taken out on the first person the chimps, now big enough to defend themselves, had a chance to punish. Moe watched helplessly as the one man who had cared for him was continuously attacked, until the other chimps were shot dead—ending their pathetic caged existences as discards of the Hollywood entertainment industry.

The media explained that Moe had been an orphan brought back by the Davis family from a vacation in Africa. The story didn't share that almost all baby chimps acquired from Africa are orphans. As with elephants and orcas, a living mother will never part willingly with her baby chimp; hunters shoot the mothers out of trees and pull the babies off their backs. For decades the babies supplied the U.S. demand arising from TV shows, circuses, and people wanting exotic pets.

"It says, no animals were actually killed in the production of this cave painting!"

Oliver's Travels

Oliver's tale teaches us about Hollywood animal lives. Oliver was captured as a baby in the jungles of what is now the Democratic Republic of Congo and sold to Janet and Frank Burger, whose animal acts were regularly featured on *The Ed Sullivan Show*. He achieved stardom in the 1970s, billed as "the Missing Link," or as a "humanzee," the billing due mostly to his humanlike two-legged walk. He also had less facial hair and what is considered to be a more human-shaped head than most chimps. He was sold by the Burgers to another trainer and toured Japan, smoking cigars, drinking whisky, and sleeping between satin sheets in fine hotels. But when interest waned, according to London's *Daily Mail*, "Oliver spent the next decade passing from trainer to trainer, appearing in jungle parks, circuses and, eventually, in roadside freak shows." Then he was sold to a vivisection laboratory, where he spent seven years living alone in a cage between painful tests.[26]

An article about Oliver in the *Atlantic Monthly* told us it was surprising that Oliver was "tractable enough" to be used in the entertainment industry into his twenties. It explained:

> A performing chimp's career is usually over by around the age of eight, though a trainer may be able to safely squeeze out a couple more years by pulling the animal's front teeth or, in the case of a male, by castration. Since chimps in captivity can live forty or even fifty years, the question arises of what to do with all those movie and circus veterans for the remaining 80 percent of their lives. Some are used to breed the next crop of performers; others end up in private homes or roadside zoos; and many, like Oliver, are sent to bio-medical research labs.[27]

That kind of makes you want to tell all those child stars who voluntarily turned to drugs that it's time to quit feeling sorry for themselves—particularly if they ever acted with chimps.

Oliver now lives at the Primarily Primates Sanctuary in Texas. Sadly, sanctuary retirement doesn't guarantee a good life. He lived entirely alone there for many years in a small cage. Now he is in poor health and almost blind. Oliver had a brief respite from solitude when seven chimps were retired to the sanctuary from an Ohio State University research program, and one of the chimps, Sarah, was housed with Oliver. People volunteering at the sanctuary, while worried about Sarah having been separated from her group, said that she had been looking out for Oliver. The two chimps had bonded and were grooming each other. Unfortunately, however, according to the *Houston Chronicle* (and reported similarly elsewhere), "Overcrowded and filthy, the facility was a squalid hoarder's camp."[28] In late 2006, after more than a decade of allegations of horrendous conditions at the sanctuary, a Travis County probate judge ordered Primarily Primates placed under court supervision.[29]

During the six-month period of court supervision, seven OSU chimps, including Sarah, were moved to Chimp Haven. Oliver was left behind. Much ado was made about his move into a much larger cage, but through most of 2007, this ex–television star, trained to sleep on satin sheets, lived in that cage alone.

The control of Primarily Primates has been returned to a restructured board of directors. Some of the sanctuary's troubles continue, but Oliver is finally doing well. An elderly female chimp, Raison, ostracized by her troop, has moved in with Oliver. Video shows them lying side by side grooming each other. With care and companionship Oliver's last few years may be as good as possible for anybody in ill health and captivity. But I hope his sad tale will help people understand why animal advocates protest the use of chimps in the entertainment industry. We are familiar with their fates.

My Dead Flicka

Not all Hollywood animals are killed behind the scenes or rot away in cages until their deaths—some die on the set. As audiences who love horses flocked to see *Flicka*, they might have wondered which of the animals they were watching were now dead. Two horses were killed during the making of that movie. Nevertheless, the film got a nod from the American Humane Association, which monitors movie sets. A *Los Angeles Daily News* article noted that the Screen Actors Guild of Hollywood funds the American Humane Association, and quoted animal advocate Kathy Riordan: "I personally think there is a major conflict of interest when the entity responsible for monitoring an industry is supported by it."[30]

When one remembers that two hundred horses were killed in the filming of the original *Ben-Hur*, *Flicka*'s awful record almost looks good. Some people have suggested that *Flicka*'s pro-mustang message makes the film worthwhile, and that the horse deaths were just unfortunate accidents. The beautiful animated film *Spirit*, however, shows us that a strong pro-mustang message can be delivered in a compelling film without the use of animal actors. It is a great rental.

Flipper's Suicide

Life in Hollywood is hard on animals, but as we saw above, life after Hollywood can be even harder; it is that which drives much animal rights opposition to the animal film business. It can even make converts of those actually in the business. When I interviewed the *Flipper* trainer Ric O'Barry for my KPFK radio show, he told me about his last moments with Kathy, the lead dolphin in the role of Flipper. He explained

that unlike other mammals, dolphins are not automatic breathers; every breath is a conscious choice, and when life becomes unbearable they can choose to take no more. They commit suicide. He says that much of the early mortality rate of dolphins in captivity is a result of suicide: "We literally bore them to death."[31]

Ric said that when the *Flipper* series ended, the dolphins were simply warehoused. Kathy was kept alone in a tank. O'Barry left for India, where he did some soul-searching, and started to feel strongly that what he had been doing to the dolphins was wrong. When he returned from India, he heard that Kathy was sick and went to see her. He found her alone in a tank, with blisters all over her back from the sun. As I interviewed him about Kathy, he got too choked up to talk about it, but on a video made by the Dolphin Project he described her final moments:

> She swam right over into my arms, looked me in the eye, took a deep breath, and never took another one. I let her go and she sank very slowly to the bottom of the tank.[32]

He describes jumping into the tank, attempting to revive her, realizing it was too late, and crying,

"My God, what have I done?"

He tells us,

Kathy was an enormously clever and bright creature, who, when there was no use for Flipper, for Kathy, had been consigned to a tank to die. And die she did, with me weeping for having done this horrible thing to her.

Dophin Project Archives

Anthony Freda

cowboys and injuries—The Rodeo

Rodeos are one of the few events still legal in Western culture where much of the cruelty happens in full public view. Perhaps most shocking is calf roping, where a terrified baby animal is pursued by cowboys on horseback, is lassoed as he runs, then is yanked off his feet by the rope around his neck, often flipping high into the air before he lands on his back. The cowboy ties up his legs. If we saw somebody do that to a puppy, we would be shocked—and might report him to the police. Perhaps because we are conditioned not to respond to the pain of species with whom we do not live, rodeo audiences seem immune to what is happening to the calf.

Shocking Abuse

Most of us grew up watching brave cowboys ride bulls and broncos for entertainment—either at the rodeo or on TV. We probably never considered that the bull would prefer to be spending his day munching in meadows and humping heifers. You might note that you don't get to do what you want to do every day either, but at least you don't have people riding on your back all day long for sport—unless maybe you are an heiress.

Events with full-grown bulls are not as shockingly abusive as those with calves—until you look behind the scenes: Just before the animals are let out of the chute, electric prods rile them up. The prods are illegal, but that doesn't always stop the cowboys or even seem to worry their fans in law enforcement. In undercover video on Steve Hindi's RodeoCruelty.com Web site, you'll see bulls and horses being shocked, seconds before their release, as local animal control officers look on from just feet away.

Some animals die in the ring. According to an article on the front page of the *Calgary Herald*, the famous Calgary Stampede loses an average of four animals per year.[34] Others go back into pens to await the next rodeo, and many, severely battered and broken, are sent to slaughter after having spent their final hours as big toys for the big boys.

Bloodless Bullfights?

There is no need to explain here why animal rights activists object to the long, drawn-out deaths of Spanish bullfights—though you can read a little more about them in chapter 8's "Running of the Nudes." People might wonder, however, why there would be objections to the bloodless bullfights that have started to make their way across the border from Mexico.

The *Miami Herald* has described the sport in gruesome detail:

> The door opens and a startled white bull charges forward. . . . In Florida's version of bull tailing, two mounted cowboys chase a bull up and down an oblong arena, competing to flip the animal over as many times as possible within two minutes. The helmet-clad rider closes in on the bull at a full gallop, grabs its tail and then leans in the opposite direction, sending the bull into a fishtail spin and tumble. . . . After one or two falls, most of the nine Brahman bulls showcased Sunday either could not, or would not, get up. . . . When the bulls refuse to stand again, out rides Cuban-American equestrian Juan Pérez Rodríguez, his white straw cowboy hat pitched low over his face, to shock them with an electrical rod. At a Southwest Miami event last year, the coleadores broke the bulls' tails, according to reports made to the Humane Society, in keeping with an old Venezuelan tradition—even though the action was officially banned by the sport's federation several years ago. "If they don't break the tail, the bull takes longer to get up," said Teresa Molinos, the sport's female world champion and a Miami resident.[35]

The article includes the following touching anecdote:

> One young bull suffered a serious leg injury after he was toppled by a horse and rider. As no veterinarian was present, the bull's injury went undiagnosed, but the startled animal could not move. He dragged his leg and stumbled, refusing to get up as the crowd jeered. Out rode Rodríguez, bearing his long electrical rod. Then, across the dusty track came the intervening scream of 10-year-old Colombian Jaime Andrés Torres: "Noooooooo!!!"

Just Kidding Fishing

Bloodless bullfights seem to be guided by the same principles as catch-and-release fishing, the idea that if the animal is not killed, what's the harm? That thinking reminds me of the Genesis Award–winning *Daily Show* report, "The Deer Hunter Pt. 1," in which Jason Jones covered "a competitive dart hunting league," who engage in what they call "nonfatal hunting." As a dart hunter describes the "humane" sport, we see a deer, hit by a dart, jump many feet into the air, then take off running faster than a speeding bullet. It hurts just to watch. Then a "real hunter" tells us that he doesn't see the point of nonfatal hunting. Jones comments, "Whatever their differences, both men agree, you just can't let animals walk around the woods unpunished."[35]

Ellen DeGeneres did a terrific skit on catch-and-release fishing. She said it is like chasing after a pedestrian in your car, swerving all over till you run him down, and then saying, as he brushes himself off, "Just wanted to see if I could hit you—you can go now."

But the fish don't just brush themselves off. A lead story in the *Los Angeles Times* on the impact of sportfishing on declining fish populations made it clear that "sportfishermen have a much larger role in depleting ocean fish than previously thought." It said, "Anglers who catch and release don't always help, because many of those fish die, the mortality increasing with depth, ac-

I don't keep what I catch. I just stick a barbed hook in its face, yank it gasping from the water, rip the hook out, & then throw it back in, leaving it to wonder what kind of god would allow such a thing.

BIZARRO.COM
Dist. by King Features

cording to the study. Pacific rockfish, for instance, when pulled from 120 feet or deeper, with eyes bugging out and air bladders expanding out of their mouths, have very poor chances of living if thrown back."[37]

It is not only deep-sea fish that die after catch-and-release. When a fishing extension agent in Key West was interviewed for a story in the *Miami Herald*, he suggested that the "assumption" is that there's a 10 to 20 percent mortality rate for fish released by recreational fishermen. The article told us, "A kind of grouper called 'gag' caught by recreational anglers has a mortality rate that is probably 80 to 90 percent."[38]

In a press release encouraging anglers to "use extra care" with catch-and-release fishing, Maryland's Department of Natural Resources Fisheries Service cited studies showing that "the physical injury from hook wounds (deep hooking) is the single highest contributor to release mortality of many species of fish." It said, "Studies by DNR's Fisheries Service have demonstrated that deep hooked striped bass will die approximately 50 percent of the time regardless of temperature, salinity or whether or not the hook is removed."[39]

Non Sequiter © 1996 Wiley Miller. Dist. by Universal Press Syndicate. All rights reserved.

Running for Their Lives—The Racing Industries

Derby Winner Dinner

In 2006, the world watched in horror as Barbaro went down at the Preakness, dashing his owners' hope for the Triple Crown. There was a public outpouring of emotion. Unbeaten in six races before the Preakness, Barbaro was potentially worth $30 million as a breeding stallion, so every effort was made to save him. He lived for eight months but was finally euthanized in early 2007—and the world grieved.

The upside of the accident was that it at least brought some attention to the plight of racehorses less famous and valuable than Barbaro. *Newsweek* reported, "If Barbaro weren't potentially worth millions of dollars, or if his owners weren't wealthy themselves, the steps he took on the track at Pimlico very likely would have been his last."[40]

The week Barbaro died, horse racing was discussed on *Larry King Live*. Jack Hanna, from the Columbus Zoo, was interviewed. We in the animal rights movement get a little perturbed when Jack gets trotted out and called an animal lover. One of the lesser-known biographical facts about that animal lover is that he appeared in a TV commercial in 1998 urging voters to oppose a ban on dove hunting in Ohio, helping to defeat the proposal at the polls. Dove hunting! On *Larry King Live* we saw him treated as an expert on all things animal. After PETA's Lisa Lange described the dark side of horse racing, Hanna represented what had happened to Barbaro as some sort of freak accident. He said he had read statistics of the numbers of animals dying years ago, but he also said, "I don't think we have those numbers dying today, especially the way these horses are cared for."[41]

Lange retorted that seven hundred to one thousand racing horses are euthanized every year. It is sad to think that the public might have been tempted to take the word of well-liked Jack Hanna over the radical gal from PETA. But in fact, just after Barbaro's accident, an Associated Press article informed us that 704 horses died while racing in 2005 in the United States and Canada—and that number did not include those who died in training.[42] As a *Philadelphia Daily News* columnist wrote, "It is not something they talk about much in their advertising, but horses die in this sport all the time—every day; every single day."[43]

Lange also noted that racehorses are given illegal and legal drugs that mask injuries. Hanna said, "As far as drugs, my understanding is these horses are checked even better than human beings like ballplayers are, before and after a game. So it's hard to believe that those kinds of drugs, as she says, are used in today's racing field." But in 2006, *USA Today* reported, "The number of racehorses that failed drug tests in California has nearly doubled since 2000, and the offenses rarely result in disqualification or other stiff penalties." It told us that California registered 142 violations in 2005.[44]

© Anthony Freda

The veterinarian Holly Cheever worked for years in the horse-racing industry. In an interview on Los Angeles's KPFK Radio she told me that horses are trained and made to race even when injured, with drugs masking the injuries, as there is a saying in the racing industry, "The horse ain't making you no money if it is standing in its stall."[45] A story on National Public Radio supported what we learned from Cheever. It said that anti-inflammatory drugs make it possible to keep training horses through exacerbating injuries, and that "few thoroughbreds in the U.S. race without medication."[46]

I take no pleasure in revealing that Hanna offered opinions as facts on topics about which he did not know enough to comment accurately—though any doves reading this book might be having a good laugh. I think it is important, however, to bring the truth to light, as Hanna often comes up against real animal advocates on talk shows, and viewers have a right to know whom they can and cannot believe.

During the KPFK interview cited above, Dr. Holly Cheever explained why horses "break down" on the track. They start training at age one, before their growth plates have closed. She said, "No sports physiologist would ever let you overrun a human athlete the way we overrun horses at such young ages." Horses that break down on the track almost always get killed immediately.

Other horses get dumped. The sheer volume of animals guarantees it. Cheever explained that to get a few dozen top-notch race foals every year, a few thousand are bred, and most of them, unsuccessful, will be discarded. Even most of the winners, almost all lame by age five or six, are discarded. She said, "There are a few good homes, but most of the horses end up bouncing from good homes to bad homes, till they end up being very neglected in someone's back lot somewhere, alone, uncared for, without veterinary care. Then I get called in by the police, and I am looking at twenty horses who are just a rack of bones full of injuries and overgrown hooves and chronic arthritic problems from their stressful life as racehorses."

Every year thousands of horses haven't gone to homes at all, they've gone straight to slaughter. Cheever said, "The breeders and trainers don't call it going to slaughter, they call it going to auction, though they know darn well there is not a huge market for half-grown thoroughbreds who really don't have any particular future. They get bought for meat."

That issue got some coverage when the story broke that the 1986 Kentucky Derby winner, Ferdinand, had died in 2001 in a Japanese slaughterhouse and had been made into pet food.

I will discuss later, in chapter 8, bills likely to pass Congress that are intended to ban horse slaughter in the United States. The American Veterinary Medical Association

(AVMA) doesn't support them, because if we outlaw horse slaughter we will still have an excess of animals—many of whom are bred for racing. The AVMA suggests that if people are not allowed to send horses to slaughter in the United States, they will be abandoned or left rotting in backyards. Their opposition points to an important fact: Legislation to ban horse slaughter is part of the puzzle, but it is not the complete solution to the woes caused by the racing industry.

After Barbaro's injury, while veterinarians fought to save his life, the *New York Times* sports columnist William C. Rhoden shared a contrasting and more typical scenario in a piece headed "An Unknown Filly Dies, and the Crowd Just Shrugs":

> There was no array of photographers at Belmont Park yesterday, no sobbing in the crowd as a badly injured superstar horse tried to stay erect on three legs. There was no national spotlight.
>
> Instead, there was death. In the seventh race at Belmont, a four-year-old filly named Lauren's Charm headed into the homestretch. As she began to fade in the mile-and-an-eighth race on the grass, her jockey, Fernando Jara, felt her struggling, pulled up and jumped off.[47]

We read of Lauren's Charm being shot and carted away. Rhoden commented,

> The scene was in stark contrast to what unfolded at Pimlico last Saturday when the Kentucky Derby winner, Barbaro, severely fractured his ankle in the opening burst of the Preakness. A national audience gasped; an armada of rescuers rushed to the scene. In the days that followed, as the struggle to keep Barbaro alive took full shape, there was an outpouring of emotion across the country and heartfelt essays about why we care so much about these animals.
>
> But I'm not so sure we do, and I'm not so sure the general public fully understands this sport.

Long Photography, Inc. Genesis Awards photograph courtesy of HSUS Hollywood Office

"People always say, the horse likes it—he wants to be broken. How do you know he likes it? What are you, f**king Mr. Ed? He told you he likes it? A bit in your mouth pulling on you. What's that, like when you first get into spanking? Oh yeah, kind of sexy. No, I am sorry, if you get on an animal's back I think he has every right to get you off his back" **(Bill Maher).**

Quote from National Animal Rights Conference, Los Angeles, 2003

J. D. Crowe/artizans.com

Dead Dog Walking

Racing greyhounds, when not on the track, spend their lives caged and muzzled—then face fates similar to or even worse than horses when their racing days are over. An HBO *Real Sports* special explained that the fastest dogs are moneymaking machines, "but the slower ones grade off, as they put it, being sent to lower and lower tracks until they can't compete anymore."[48]

In that piece, a track worker says that in the greyhound business, a dog who had lost too many races would be described, coming off the track, as a "dead dog walking."

The lucky ones are adopted out, but more are killed. In 2002, a mass grave was discovered in Alabama holding the bodies of three thousand greyhounds who had been shot for $10 each in a kill-for-hire operation that serviced Florida's dog-racing industry.

A piece in the UK's *Independent* personalized the issue. It opened:

> Rusty the greyhound's toe injury proved to be fatal.
>
> After performing poorly during a race at Warwick in April, the once-prized sprinter could no longer earn its keep. The following week, Rusty was discovered by a walker in South Wales,

lying whimpering on a rubbish tip, its tail still wagging. The dog had been shot through the head with a captive-bolt pistol, its ears cut off to remove identifying tattoos. A vet was called to finish the bungled job of killing the dog.[49]

As with Hollywood castoffs, the most unfortunate retirees from the greyhound racing industry are dumped into vivisection laboratories.

Photograph compliments of David Duchovny

David Duchovny, pictured here holding his lovely Delilah, says, "Skip the racing track. Adopt a greyhound and let her race along the beach with the rest of your rescues."

The Great Turtle Race

Many of us who live with animals have noticed that they love to race. Buster and Paula used to pull me on Rollerblades up New York's Lafayette Street, in the bike lane. Whenever a bicycle came up beside us, the dogs, with no encouragement from me, would double their speed, taking side glances at the bike rider, and only relax to their normal pace once the bike was about a block ahead, well out of catch-up range. They seemed to be having a blast. What a shame that the natural and joyous behavior of animals becomes deadly when human commerce comes in. While humans love to bet, the animals all eventually lose.

Occasionally those natural racing and betting instincts are parlayed into something good: In the Great Turtle Race, eleven endangered female leatherback sea turtles, each with her own nickname, sponsor, and satellite tracker, make their way from a Costa Rican beach to the Galapagos Islands.[50] Each turtle has an adorable baseball-like card complete with vital stats. Stanford University's Mark Breier and George Schillinger cofounded the race after learning of an event involving albatrosses, sponsored by a British big-bird conservation group. Of the turtle race, Schillinger said, "It's like the Tour de France or like the Chinese Olympics." But its goal is to raise awareness and concern for the turtles' potential extinction—and to raise money for conservation efforts. The founders did a great job of raising awareness by naming one of the turtles after the TV host Stephen Colbert. With much pride, Colbert covered the race on his show, and naturally put his money on Stephanie Colburtle the Turtle. Colbert showed his audience a photo of his "beautiful adopted daughter" and triumphed, "Eat it, Angelina!"[51] He boasted that she is the largest of the turtles in the race, explaining it isn't her fault—she is just big-shelled. You can learn more about the event at www.greatturtlerace.com. It shows us that we can have fun betting on animals, while helping rather than hurting them.

A better tomorrowland

Ric O'Barry says that the most important thing people can do to discourage the cruelty of the animal entertainment industry is to stop buying tickets. Would that mean the end of circuses and zoos? Will people never get to view wild animals? And will there be no more animals in the movies?

Curtailing the use of wild animals in entertainment certainly wouldn't mean the end of circuses. Circuses that rely on human performers, such as the glorious and immensely popular Cirque du Soleil, point the way to the future of circus entertainment. There are now more than twenty animal-free circuses around the United States.

Though some argue against it, I feel comfortable with the continued use of dogs in the entertainment business, whether it be in live performances or in Hollywood. If rescued from shelters, as was the case, for example, with the stars of the *Benji* movies, dogs can be offered life by the entertainment industry, rather than death by lethal injection. And dogs are pack animals—kind of like the guys on HBO's *Entourage*. Can you imagine Turtle's pitiful whining if Vince left him home alone all day? He would argue that Holly-

They train me to perform, then when I try to show off what I really do best, everybody goes ballistic.

COURT-ORDERED PSYCHOLOGICAL EVALUATION

Bizarro © Dan Piraro/King Features Syndicate

wood dog lives are better than the lives of millions of dogs who spend every day alone in backyards while their human families are out in the world. Sure, dogs used in entertainment are not totally free to determine their own lives. But most humans go to work every day at jobs they don't adore when they would rather be lounging on the beach. It is not a tragedy—and when we focus on it we risk crying wolf and distracting attention from the real tragedies. If we protest the use of clearly content dogs in films, we may not be taken seriously when we explain that wild animals, such as elephants, accept their tragic entertainment-industry lives only because they were beaten into submission and taught, while they were still babies, that protest was futile.

Shutting down the use of wild animals in entertainment wouldn't mean we would never see them in movies. Recent breakthroughs in computer-generated imagery allowed for the thrilling scenes of tiger fights in *Gladiator*, without the employment of animal actors. The digital cow hit by a car in *O Brother, Where Art Thou?* provided gruesome entertainment with no suffering either on set or behind the scenes. And the movie *Greystoke* showed us what human actors in suits can achieve when they learn to act like chimpanzees. Dr. Roger Fouts, who was a consultant on the film, points out that the human actors in chimp suits got paid—and they didn't get sold for biomedical research when the filming was over. More recently, with the use of terrific actors and special effects, *King Kong* director Peter Jackson made sure that movie's only tragic tale of a captive ape was the one being told on the screen.

As for ever viewing real wild animals, sanctuaries, such as Carol Buckley's in Tennessee and the PAWS sanctuary in Northern California, do not allow visitors. The sanctuary owners feel that the animals have served the public for long enough and that we now owe them a haven free from entertainment duties. That is an understandable position. But what about once we have dispensed with abusive animal circuses, and with zoos that trade animals with no thought for their familial bonds and individual needs? Could we not imagine a compromise, where the needs of the animals are considered first and foremost, but where people are allowed viewing privileges?

We already have similar places that could serve as models—the wonderful farm animal sanctuaries across America, where one can spend the day hanging out with rescued farm animals.

In January 2006, it was announced that Ngamba Island (near Entebbe, Uganda), which was set up in 1998 as a sanctuary for chimpanzees confiscated from smugglers, had been opened to the public and was offering half-day, full-day, and overnight viewing trips. Perhaps that type of place could serve as a perfect model for parks that could eventually replace the zoos that currently put the needs of the animals well behind the needs of the patrons looking for a day's entertainment.

"We had to let the animals go. No one informed them
of their rights when they were arrested."

I agree with Ric O'Barry, who says that viewing wild animals in cages and tanks teaches people only lessons in domination—they learn that it is okay to mistreat animals for human amusement. But perhaps if people drove through wild animal sanctuaries, and saw animals in family groups engaging in their natural behaviors, it would be truly educational. Perhaps it would teach a real appreciation for the magnificence of other species and be beneficial for all. Those sanctuaries might even have breeding programs, so that species do not disappear, the breeding occurring either within the sanctuaries or, only if it is deemed best for the animals, by artificial insemination, thereby increasing genetic diversity without shunting animals across the world.

Above, I described the plight of the orca Lolita, whose family awaits her in Puget Sound. If she were released back to them we might see a reunion, also after thirty years, like that of the elephants Shirley and Jenny. Yet there are dolphins and other whales who are injured, or who have never known life in the ocean, for whom a release into the wild is not an option. Surely we can do better for them than concrete tanks. O'Barry suggests we net and fence off inlets or bays, where animals who cannot be rehabilitated would be able to swim for long distances in straight lines, could feel the motion of the tides, could enjoy chasing fish, and would never be forced to perform tricks to amuse humans.

It is not unusual for wild dolphins to actually seek out humans. As I look out over the southern California beach near my home, I see people kayaking with dolphins swimming next to them. Dolphins also gravitate to sailboats, and even to motorboats, seeming to enjoy playing in the waves the vessels create. Scott Olson, I hear, takes people on tours to swim with wild dolphins in Hawaii. He cannot guarantee a meeting, because the dolphins come or not as they please. But he tells me that anybody who has come for a few days has never been disappointed—the dolphins, when spotted, are generally happy to interact with the swimmers for a while. So are sharks, but with different results. (I kid about that only so I can point you to "A Shark's Tale" in chapter 5, to read about how much more sharks have to fear from us than we do from them.) Going out to sea in search of dolphins is not nearly as convenient as jumping into a tank with captive dolphins. But how wonderful to interact with animals who have not been taken from their homes and families, who have not been forced to interact in order to be fed.

Such solutions are not always perfect. Some whale-watching trips, for example, have been reported to disturb the whales. Granted, not as much as harpoons. But boat engine noise can seriously interfere with their ability to navigate. Such issues, however, can surely be overcome so that we can find ways to enjoy viewing and interacting with other animals, while keeping the focus on their safety, well-being, and right to freedom.

There are animal rights activists who say the only appropriate thing to do with wild animals is to leave them alone—to have no interaction. But as their habitats disappear, in ways we cannot always control, that absolutism might not always be kind. At the other extreme are people who feel the animals are on earth for us to use as we please and that dominion should mean domination. They see no problem with the animal entertainment industry and suggest that those animals in that industry, or in zoos, are happy to be ambassadors for their species. When Jack Hanna tried that one on Bill Maher, Maher's exasperated response was, "Did they tell you that, Jack? Is that what the marmoset said: 'I'm an ambassador!'"[52] Bill wasn't buying that the marmoset might have some interest in diplomatic immunity or in unrestricted parking without tickets.

If Jack stopped parading animals around on stages long enough to listen to them, perhaps he would hear that marmoset, and hear all the whales at the SeaWorlds he promotes, saying "I want to go home."

Bizarro © Dan Piraro/King Features Syndicate

chapter four

fashion victims: animal clothing

Look at what some of us put ourselves through to look good—starving, purging, jogging till we drop, breast implants when we are young, and all sorts of lifts when we are old. Is it therefore surprising that not too much attention is paid to the animals who suffer so we can look good? Yes. A lot of people would say you are free to hurt yourself to your heart's content, but don't hurt anybody else. While it is no secret that wearing fur hurts somebody else, most people probably don't know how much. And the pain associated with other animal fabrics is an even better-kept secret.

They say "you can't take it with you," but guess what—the animals & I decided to let you keep all those furs you owned!

Bizarro © Dan Piraro/King Features Syndicate

Fur—you have to be pretty cold

To make a forty-inch fur coat it takes between thirty and two hundred chinchilla, or sixty mink, fifty sables, fifty muskrats, forty-five opossums, forty raccoons, thirty-five rabbits, twenty foxes, twenty otters, eighteen lynx, sixteen coyotes, fifteen beavers, or eight seals.

Most animals killed for fur come from fur farms—hideous places where the animals are kept in tiny cages. They are generally killed at approximately six months of age, but the females used as breeders are housed that way for years. In nature they would build burrows in which they would huddle together for warmth, but on fur farms they are often left exposed to the elements in order to encourage the thickest coats possible. When animals freeze to death—no worries—they can still be skinned.

Fur ranchers choose the cheapest, not the kindest, kill methods, the main concern being the preservation of the whole coat. Many ranchers use anal electrocution, which fries the animal from the inside out, microwave style. At FurIsDead.com there is footage of animals undergoing anal electrocution, including an unsuccessful attempt repeated on the same animal. We also see a farmer who injects insecticide into the chests of minks, who die convulsing in agony over many minutes.

The fur industry eases consciences by stressing that most fur these days comes from farms. (Didn't the above descriptions make you feel better?) But millions of animals are still trapped for their fur. The steel-jaw trap is popular. Animals are often left alive for days, in excruciating pain and exposed

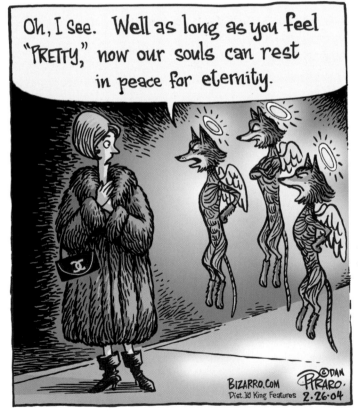

to the elements. Mothers desperate to get back to their young have been known to chew off their caught limbs. To protect the valuable fur from mutilation by predators, pole traps are often used. Animals caught in pole traps are hoisted into the air and left to hang by the caught limb until they die or until the trapper arrives, often days later, to kill them. All for a trendy coat collar.

The *New York Times* ran a story on a recent hot trend in fur—astrakhan, also known as karakul or broadtail. The article explains that astrakhan lambs are killed at just a few days of age, "and some examples, called broadtail, often considered the most desirable, are the skins of unborn lambs."[1] The mother's throat is slit and her stomach slashed open to remove the developing lamb. Boy, you sure have to be pro-abortion to go for that one.

The world's number-one fur producer is now China. At FurIsDead.com you can see undercover footage taken at fur farms there. Animals are slammed against a concrete floor, then skinned while stunned but still conscious and blinking or even struggling. Some, once skinned bare, die slowly over the next twenty minutes.

Going Clubbing

Every year hundreds of thousands of seal pups, over twelve days of age but still too young to swim away, are clubbed to death on the Canadian ice floes. An op-ed in the *Christian Science Monitor* tells us,

Did you or did you not know there were hunters in the area when you placed the realistically painted, cast-iron baby seals on the ice?

BIZARRO.COM
Dist. by King Features 3·29·03

Karremann/PETA

Karremann/PETA

Karremann/PETA

Frank Micelotta/Getty Images

"There are no regulations governing the Chinese fur industry, which means a miserable life and an excruciating death for every animal. If you buy or wear fur, you must share the blame for the suffering of these cats and dogs" (**Trent Reznor**, Nine Inch Nails).

Video PSA available online at www.furisdead.com/feat-trentreznor.asp.

In 2001, an independent team of veterinarians was escorted to the ice floes by the International Fund for Animal Welfare. They studied Canada's commercial seal hunt at close range. Their report concluded that up to 42 percent of the seals they studied had probably been skinned alive while conscious—a clear violation of Canada's criminal code and marine mammal regulations that govern the hunt.[2]

There have been many international protests.

In October 2006, the German parliament banned the import of all seal products. Belgium has followed suit. Members of the European Parliament have passed a non-binding declaration calling for a ban on imports of seal products—a declaration that may become law. But for now people still wear sealskin coats and the hunt continues.

At protectseals.org you can see video—horror films—of the Canadian seal slaughter and follow the latest developments.

How Much Is That Dead Doggie in the Window?

CNN's *Larry King Live* aired undercover footage from the lucrative Chinese trade in dog and cat fur. Viewers saw close-ups of crates full of dogs, who were dropped from the tops of trucks to the road below, shrieking as they crashed against the pavement, breaking their bones. A German shepherd was tied up just before his inner thigh was gashed and he was bled to death. (The show chose not to air the actual cutting and bleeding, but you'll find it at FurIsDead.com.) We saw two cats clinging together, with their forearms around each other's bodies, cowering at the back of a crate. Rick Swain, from the Humane Society of the United States (HSUS), explained that the crate had held fifteen cats. The two had watched all the other cats being pulled to the top of the crate by a noose around the neck and strangled there. We saw one of the two pulled away as the cats frantically tried to clutch each other. We also heard descriptions of dogs being skinned alive and left to die.

Dog and cat fur, cheap and plentiful, is sold throughout Europe under many different misnomers to disguise its origin from Westerners who care about dogs and cats. Dog fur might be labeled Mongolian Wolf. The sale of dog and cat fur has been banned in the United States, but imported coats are not DNA tested; we rely on labels. On *Larry King Live* we heard the following from HSUS's Rick Swain:

> When I was in China, I was accompanied, as you probably know, in many towns by a government minister. I saw a well-known American label on a coat hanging on a rack. I knew it was dog fur and I asked about it. I said, "What's with this label?" And the minister just laughed and said, "This is China, we'll put any label on it you want."[3]

A Fox 5 Washington, D.C., news report publicized tests in which the Humane Society of the United States subjected twenty-five fur-trim jackets from twenty different retailers to mass spectrometry testing; they found that all of the coats were mislabeled. In most cases they were labeled as having fur from a legal animal, such as a raccoon, when in fact the fur came from a raccoon dog, which looks a little like a raccoon but is part of the dog family.[4]

Faux Faux

Many of the coats tested in the study cited above were labeled as faux fur although real fur had been used. The Fox News piece even reported that a labeling loophole allows coats with less than $150 worth of fur not to be labeled as including fur at all. They explained that means a coat could get through without a fur label even if it included the skins of thirty rabbits, or three raccoons, three red foxes, or one bear. Or, as indicated by the HSUS study, three dogs.

Do Nice Girls Fake It?

Even when the fur really is fake, there are problems, as exemplified in an article in the *Boston Globe* about Martha Stewart:

> When Stewart walked out of the federal court-house—and in front of a bank of cameras—with a furry accessory knotted cozily around her neck, her fashion statement set off two reactions: The first was from the People for the Ethical Treatment of Animals, which promptly named her one of the world's worst-dressed celebrities of 2004. The second was from furriers who raced to emulate what they believed was Stewart's dyed chinchilla scarf. Both were misguided. Stewart was wearing a fake.[5]

"Faux fur? Oh, well, that's faux blood."

www.cartoonstock.com

Martha's fur scarf being fake made a difference to the animals who would have died for her scarf if it were real, but made no difference to the thousands of others who were killed thanks to her inadvertent example. Stewart has since made efforts to balance the books by recording an advertisement for People for the Ethical Treatment of Animals in which she says, "I used to wear real fur but, like many others, I had a change of heart when I learned what actually happens to the animals."

Some animal rights activists would argue that the skins of tortured animals draped around a human, and garments fashioned to resemble that horror, are similarly grotesque.

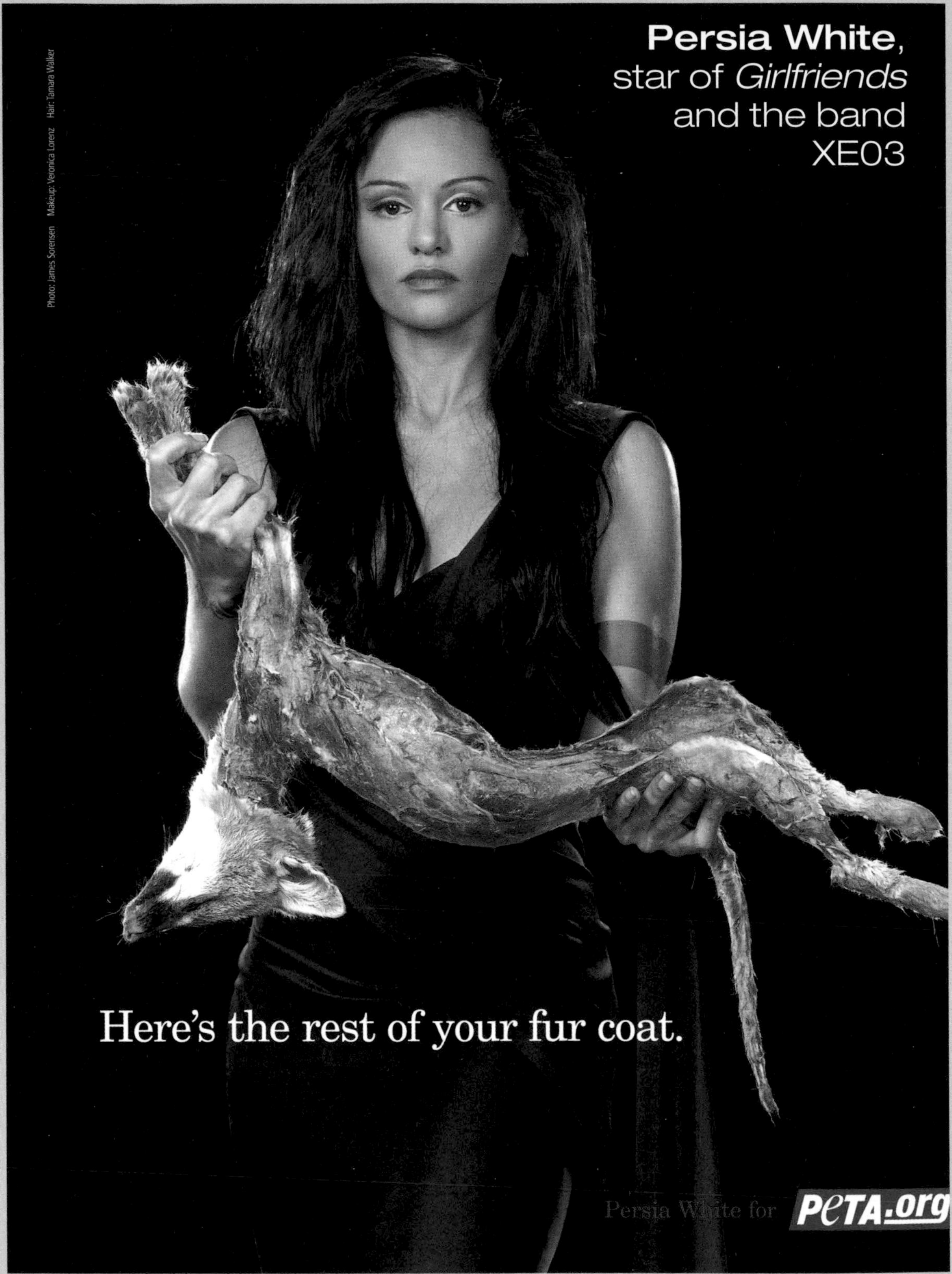

Persia White,
star of *Girlfriends*
and the band
XE03
Photo: James Sorensen Makeup: Veronica Lorenz Hair: Tamara Walker
Here's the rest of your fur coat.
Persia White for PeTA.org

"Some of the practices are so cruel, and as a celebrity you have a responsibility to think about the message you're sending out by wearing fur. People will think it's okay or cool, but it's not" (**Pink**, on celebrities who wear fur).

"Pink Slams Fur-Wearing Beyonce," WENN Entertainment News Wire Service, November 12, 2006.

Dave Hogan/Getty Images

Leather—A Fatal Fetish

Those who would not wear fur, but who eat beef, are often comfortable wearing leather, believing it to be a beef by-product. That is not always the case. In India, where cows are considered sacred and cannot be killed, they are taken on endless death marches across the country so that they can be slaughtered outside Indian borders. Leather is the main product, not the by-product. A PETA undercover investigation of the trade describes cows being marched for hundreds of miles through parched deserts. Their handlers spray mace in their eyes, whip them, and twist and break their tails to get them up and moving when they collapse. The cows are often attached to each other with ropes through rings in their sensitive noses.[6]

The investigation led to a protest against the Gap clothing chain's use of Indian leather. That company agreed to discontinue its use, but the sale of Indian leather is still common throughout the world.

In Western nations leather generally comes from cows who have been slaughtered for meat. Yes, it is a by-product, but one from an animal who has lived a miserable life on a factory farm and may have been hacked to pieces while conscious in a poorly monitored slaughterhouse. (See the section "Dying Piece by Piece" in chapter 5.) Vegetarians who might consider wearing leather, because it is only a by-product, should consider that leather is the most economically valuable by-product of the meatpacking industry. The sale of leather goods bolsters the economic success of slaughterhouses and dairy farms.

Yeah, he's a nice enough guy, I guess, but that stuff his boots are made of seems eerily familiar.

Bizarro © Dan Piraro/King Features Syndicate

While leather producers often tout their product as more eco-friendly than synthetics, most leather is chrome-tanned, producing waste deemed hazardous by the Environmental Protection Agency. The groundwater near tanneries can hold highly elevated levels of lead, cyanide, and formaldehyde. The Centers for Disease Control and Prevention found that the incidence of leukemia among residents in an area surrounding one tannery in Kentucky was five times the national average.[7]

The 1998 movie *A Civil Action*, based on the book by Jonathan Harr, dramatizes a similar situation: After the deaths of eight children are linked to polluted water, residents of a town in Massachusetts successfully sue a tannery. Renting that film might be a fun way to learn about, or even educate your environmentally conscious family about, the impact of the leather industry.

Animals not used for food are also hunted or farmed for leather. The *Los Angeles Times* ran a story about a Georgia farmer who kept ten thousand miserable alligators, destined to become shoes and wallets, in four buildings, with hundreds of alligators crammed into every room.[8]

"It's the ghost of that pocketbook I gave you for Christmas."

Christophe Simon/Getty Images

Spider-Man actor **Tobey Maguire** has given up leather. He has also been vegetarian for fourteen years and doesn't eat meat, eggs, cheese, or milk. Maguire says, "I don't judge people who eat meat, that's not for me to say, but the whole thing just sort of bums me out."

"Tobey Maguire's Leather Ban," *People*—Monsters and Critics, April 10, 2007.

Not so warm and fuzzy wool

Sheep used for wool live mostly outdoors and are shorn at the end of winter when their coats are at their fullest, before they would naturally shed. Post-shearing, Australia loses about one million sheep every year to exposure. A few lost animals do not affect each farmer's overall profitability.

Some luxury wool, however, comes from sheep kept permanently indoors so that the wool stays soft—factory farming for fabric.[9] Similarly, other animals, such as rabbits bred for angora, are factory farmed and live cramped in small cages.

Like animals raised for food, sheep raised for wool are castrated and have their tails docked—without anesthetic.

Sue Coe on mulesing.

Prince won a Genesis Award for the following words on the *Rave un2 the Joy Fantastic* liner notes:

If this jacket were real wool, it would have taken 7 lambs whose lives would have began like this...

Within weeks of their birth, their ears would have been hole-punched, their tails chopped off, and the males would have been castrated while fully conscious. Xtremely high rates of mortality r considered normal: 20 2 40% of lambs die b4 the age of 8 weeks, 8 million mature sheep die every year from disease, xposure, or neglect. Many people believe shearing helps animals who would otherwise b 2 hot. But, in order 2 avoid losing any wool, ranchers shear sheep b4 they would naturally shed their winter coats, resulting in millions of sheep deaths from xposure 2 the cold. Respect all of God's creatures.

"2 my mind, the life of a lamb is no less precious than that of a human being." Mohandas Gandhi

CHERISH THE GIFT OF LIFE & RAVE UN2 THE JOY FANTASTIC

As **Prince** accepted his Genesis Award, he gave the ceremony the following beautiful endorsement: "I've received many awards, but I'm truly humbled just to be here. . . . There are so many wonderful people in the room. . . . This has been a moving experience just being in a room with all of you."

Fourteenth Annual Genesis Awards, Beverly Hilton, March 18, 2000.

A Pound of Flesh

In Australia, which supplies approximately 25 percent of the world's wool, sheep are bred to have skin with many folds, allowing each animal to hold more wool. The folds attract maggots, particularly around the anus. In a condition called flystrike, the maggots eat the sheep alive. The traditional solution has been a practice called mulesing. Farmers slice off hunks of flesh from the lambs' hindquarters so that smooth scar tissue forms. We all know that women in Beverly Hills pay for similar procedures, but the lambs have no say in the matter, and, most importantly, no anesthesia. Although pesticides can beat flystrike, they have to be reapplied regularly, meaning more labor and cost—so farmers generally don't bother with them. Under international pressure and boycotts, Australian farmers have agreed to phase out the barbaric practice of mulesing by 2010. At least until then, much Australian wool will come from mulesed animals.

Sheep Shipping

The wool trade supplies the live export trade, in which millions of elderly sheep are shipped every year, in horrifying conditions, from Australia to Islamic countries. There the sheep are killed by ritual slaughter using means illegal under Australia's humane slaughter laws. Widespread attention was drawn to the cruelty of live export when in 2005 a shipload of fifty-three thousand live Australian sheep, four thousand more already having died on the journey, was rejected by Saudi Arabia. The sheep remained at sea, dying slowly on the sweltering ship, from August 5 till late October. While Australian activists had protested the live export business for years, that event brought international attention to the issue, even making the front page of the *Los Angeles Times*.[10]

Unfortunately, even that shocking event did not permanently bring live export to a halt. Buying Australian wool still supports that cruelty. Activists are hopeful that will change over the next few years.

DOWN—No comforter for the animals

Much down comes from animals slaughtered for food. It adds to the profitability of animal flesh industries, such as the hideous foie gras industry. In the United States, birds are exempt from humane slaughter laws (See "Fowl Outs" in chapter 5), so the animals under whose down we have been snuggling up in our beds have died painfully. Most have also lived painfully under factory-farm conditions.

Geese and ducks from breeding flocks, who live for years, may suffer the added pain of feather removal every six weeks throughout their lives. Regular feather plucking of live conscious birds is common in Poland and Eastern Europe. Those of us who have undergone any kind of body waxing have some idea of how that must feel. Those who haven't, but have seen *The 40-Year-Old Virgin*, saw Steve Carell's character bolt out of a body-waxing session after enduring just a couple of strips. He was willing to let his unfinished chest look like a half-mown lawn rather than endure the agony of continuing the wax.

The scene is hilarious in the context of the film, but it is not funny to think of uncomprehending little creatures being put through that torture because we think their coats feel nice on us.

"What's in the coat, Jack?"

Karen Dawn

silkworm union?

No, we don't want to get silkworms a union. But we would rather not have them killed by the thousands to decorate our necks. Even Donald Trump, so passionate about silk ties and capital punishment, probably had no idea the two were related. We know he wouldn't want to see anybody killed for a job well done. And when the job was poor, surely he never meant "You're fired!" literally. But that's the thanks the silkworms get for your swanky silk shirt. The cocoons are flash heated with the larvae still inside. Approximately three thousand worms are killed for each pound of silk. It isn't necessary for them to be killed—we can wait for them to mature and fly away. But killing them is fastest and cheapest, so that's what's done in our society, where animal life is considered worthless.

For anybody who would carry an insect outside of her house rather than killing without cause, silk is an inconsistent choice. Some people, however, think nothing of killing insects, perhaps believing them to be without intelligence or sentience. They should enjoy the section on honey, in chapter 5, which presents startling information on the intelligence of bees. A *Reader's Digest* article even reported, "The humble honeybee possesses the second-most complex language, after humans."[11] According to the article, bees can send ten thousand distinct messages. And that's without ever spending a penny on an iPhone.

A *New York Times* article tells us about various studies on personality in other insects. Jason Watters, a behavioral ecologist at the University of California who studies water bugs, discusses their different but consistent behavioral types. He says, "In the presence of a predator some individuals will run and get right out of the water. Others don't seem concerned whatsoever. Just sit there."[12] We might call them Bruce Willis–types—especially the "consistent" part. It will be *Die Hard 25* before Bruce runs away from trouble. Then there are the Mike Tyson water bugs: Watters writes, "We've found among the males that there is the consistently more aggressive guy—so that's his type or his personality—and then there are these very active, hyperaggressive males. They're the ones who are always forcing females to have sex and driving them out of the water and really messing things up for themselves and everybody."

The article also discusses research on fruit flies. Apparently, some are as shy as Shrek and others party like Paritney. We learn that one researcher and her students "encountered everything from overly shy, timorous fruit flies to bold trailblazers to downright feisty and ultimately self-defeating bullies."

Spiders are closely related to insects and also produce a silk of sorts. Recent studies on them are enough to make one wonder about *Charlotte's Web*. An *Asian News International* article quotes Linda Rayor, an entomologist with Cornell University, who says of spiders: "They are constantly exploring one another and interacting with their siblings."[13] Describing her work with spiders from the arachnid class called the *Amblypygids*, it tells us:

> The two Arachnid families were often seen engaging in sibling-sibling and mother-baby interactions. In one experiment, the siblings were removed from a familiar cage and placed randomly into a large unfamiliar cage. Within minutes, they gathered back together.
>
> Mothers of both species nurtured their young. Often, the mama whip spider would sit in the middle of her offspring and slowly stroke their bodies and whips with her own feelers.

Then there are cockroaches. I have heard that if a woman in New York wants to get the cockroaches out of her apartment, all she has to do is ask them for a commitment. Well, they might not really be that much like men, but an article in the online journal *Public Library of Science* reported that cockroaches have a memory and can be taught to salivate in response to neutral stimuli—like Pavlov's trained dogs.[14] Maybe they are that much like men.

We have yet to see studies on the personalities of silkworms. It seems unlikely, however, that spiders would show nurturing behavior and familial recognition, and that some insects would exhibit intelligence and personality traits, while other insects would be simple automata. And so, giving insects the benefit of the rapidly disappearing doubt, those concerned with kindness to animals might choose fibers other than silk.

Ahimsa silk, produced in India for Hindus, is made from the cocoons of caterpillars who have made it to the moth stage and flown away. Its use for clothing seems to be one of the very rare examples of a truly symbiotic relationship between humans and other animals. The product is still rare in the United States, but that could change with growing demand. A *New York Times* article titled "Uncruel Beauty" tells us:

> Jack McKeever, a singer, musician and sometime vegetarian, stopped by Organic Avenue last week partly because he was hoping to buy an ahimsa silk suit that had earlier caught his eye. As he admired the fabric and the look, he said. "If these people can compete aesthetically, I say, 'Rock on!'"[15]

Toward Fashion with Compassion

Do we help the animals by throwing away leather shoes we already own? Unfortunately, if we wear leather shoes that are attractive we advertise a cruel industry. Many new vegans therefore choose to donate their leather items to charity. Each pair donated goes to a person with less choice, and it may take the place of another leather pair that person would have bought retail.

Another argument against wearing leather shoes we already own is that it makes us look inconsistent. Those who wish to poke holes in vegan philosophy look for those holes in our leather shoes.

Glaring inconsistency is less of an issue with wool and silk because few people are aware that there is cruelty in those industries. Many of us who would not support the wool or silk industry with new purchases still wear the wool and silk we have owned for years. On the rare occasions we are asked about our inconsistent clothing choices we can explain our stance and say that we would no longer buy those fabrics. We look forward to the day when people are so aware of the cruelty behind wool and silk that we feel forced to donate that which we own.

Because fur is a controversial issue, wearing it amounts to taking a stance—a stance against the animals. So those who care about animals will generally not even wear fur from animals killed long ago. Those old fur coats, however, can be put to good use. Fur donated to humane societies can be used as comforting bedding for homeless animals. Wildlife rehabilitation facilities put donated fur in the beds of animal orphans.

PETA uses donated coats in demonstrations against fur wearing. The organization also donates fur coats to homeless people, knowing they don't have the choices many of us have. Outfitting the homeless in fur, while helping humans in need, also helps remove some of the item's cachet. I think it is ethically acceptable to outfit in fur those in danger of freezing to death. Matthew Scully sums up the difference beautifully in his book *Dominion*:

> For ages people needed furs to survive in the severe elements we faced. Women who today keep the fur industry thriving, in order to be seen swathed in mink on a 60-degree December evening in Beverly Hills, or in Manhattan making the harsh winter trek from Saks to Tiffany's, do not have the same excuse.[16]

When giving up attire made from animal products, it is not a matter of choosing animals over people, but rather of choosing the animals' interest in life itself over our interest in every whim of fashion. We can look absolutely fabulous, or as my stylish MTV producer friend Joshua Katcher likes to say, "ethically fabulous," without wearing animals.

Anthony Freda

chapter five

Deconstructing Dinner

at the slaughterhouse

Dying Piece by Piece

According to Human Rights Watch, the most dangerous job in America is slaughterhouse work—and they mean for humans.[1] Production-line speeds at slaughterhouses have doubled in the past twenty years. The "bumper," the person with the stun gun, cannot always keep up. Improperly stunned animals contribute to worker injury rates as the animals struggle and kick.

A *Washington Post* front-page story titled "They Die Piece by Piece" describes cows bellowing and looking around in panic as they make their way down the slaughter line. And it describes, as follows, undercover video taken at the IBP slaughterhouse:

> Some cattle, dangling by a leg from the plant's overhead chain, twist and arch their backs as though trying to right themselves. Close-ups show blinking reflexes, an unmistakable sign of a conscious brain, according to guidelines approved by the American Meat Institute.[2]

The article refers to humane slaughter law enforcement so lax that no action was taken against a Texas beef company that was cited twenty-two times in one year for violations that included chopping hooves off live cattle. A technician from a Florida plant gives the following quote:

> I complained to everyone—I said, "Look, they're skinning live cows in there." Always it was the same answer: "We know it's true. But there's nothing we can do about it."

The *Washington Post* article also includes an account of the treatment of pigs:

> Hogs, unlike cattle, are dunked in tanks of hot water after they are stunned to soften the hides for skinning. As a result, a botched slaughter condemns some hogs to being scalded and drowned. Secret videotape from an Iowa pork plant shows hogs squealing and kicking as they are being lowered into the water.

Temple Grandin, an expert on humane slaughter, contends that there has been some improvement in the last few years. In 2003, she noted that 94 percent of slaughterhouses stunned 95 percent of the animals on the first attempt, meeting the legal requirement.[3] But that ratio still means that many thousands of animals experience prolonged and painful deaths at slaughterhouses every day.

Eric Schlosser, the author of *Fast Food Nation*, writes that he watches "the knocker knock cattle for a couple of minutes"—just a couple of minutes—then, "a steer slips from its chain, falls to the ground, and gets its head caught in one end of a conveyer belt. The production line stops as workers struggle to free the steer, stunned but alive, from the machinery."[4]

In a *New York Times Magazine* article on grass-fed beef (which I will discuss further in the section "This Steer's Life"), Michael Pollan tells us that after a steer is stunned, "a worker wraps a chain around his foot and hooks it to an overhead trolley. Hanging upside down by one leg, he's carried by the trolley into the bleeding area, where the bleeder cuts his throat."[5] Pollan has been assured by Temple Grandin that "they have another hand stunner in the bleed area" for animals that have been missed. Is it supposed to be comforting to learn about the bleed-room stunner, knowing that legally 5 percent of these huge animals go fully conscious all the way to the bleed room hanging from the trolley by one leg? Imagine the agony a fully conscious twelve-hundred-pound animal would experience being hoisted up, and

hanging on a moving trolley, by one leg. Anybody who has ever felt the searing pain of a dislocated shoulder might have a clue as to what those animals feel before they are killed.

Don't forget, the IBP undercover footage has made it clear that even the second stunner doesn't always get the job done.

Fowl Outs

Improved enforcement is irrelevant to more than 90 percent of animals slaughtered every year in America; turkeys and chickens are exempt from federal humane slaughter laws. When they arrive at the slaughterhouse, chickens are dumped from crates, falling several feet onto a conveyor belt, often breaking limbs as they land.

Though the Humane Slaughter Act prohibits shackling and hanging conscious animals by their legs (though it allows for a 5 percent error rate), it doesn't count birds as animals. Workers jam their feet into trolleys. The trolleys carry them along upside down and dip the animals' heads in water charged with enough electricity to leave them paralyzed but conscious, ensuring that their hearts keep beating through most of the slaughter process and they "bleed out" efficiently. After the electric water bath, a machine cuts the paralyzed birds' throats. Undercover slaughterhouse video shows that the current is not always strong enough to even paralyze the animals. Frequently the blade misses because the birds are trying to right themselves. Records acquired under the Freedom of Information Act tell us that millions of chickens every year enter the defeathering scalding tanks while still alive.[6]

Bizarro © Dan Piraro/King Features Syndicate

An article in *Gourmet* magazine, not exactly an animal rights rag, tells us:

> The chicken council allows a rate of up to 2 percent for such incidents—which means that up to 180 million chickens each year suffer through a botched death in the slaughterhouse.[7]

Chickens missed and boiled alive are referred to in the industry as "redskins." Why leave a word with only one offensive connotation if it can have two?

The Humane Society of the United States is currently suing the USDA in an attempt to add coverage of turkeys and chickens to the Federal Humane Slaughter Act. For now, however, birds remain exempt.

Egg-laying hens, stuffed into tiny little cages throughout their lives, also have hideous deaths. If they make it to the slaughterhouse they are killed like other birds. The meat left on their spent bodies, however, is often not deemed worth the cost of sending them to market. We will look at the unconscionable ways in which they may be discarded in the section on eggs later in this chapter.

Pilgrim's Shame

Not only are standard slaughtering practices inherently abusive, too often reports come out of slaughterhouses detailing intentional abuse heaped on defenseless animals by frustrated workers. People for the Ethical Treatment of Animals released undercover videotape taken at the Pilgrim's Pride slaughterhouse, which supplies KFC restaurants. The videotape shows workers kicking and stomping on chickens and smashing them against walls. PETA also supplied eyewitness testimony telling of employees "ripping birds' beaks off, spray-painting their faces, twisting their heads off, spitting tobacco into their mouths and eyes and breaking them in half—all while the birds are still alive." When the undercover report was widely aired to shocked viewers on news broadcasts throughout the United States,[8] several workers were fired but none were prosecuted.

Butterball Bullies

PETA also conducted an undercover investigation and shot footage at a Butterball turkey slaughterhouse in Arkansas.[9] Butterball workers were documented punching and stomping on live turkeys and slamming them against walls. PETA reports, "One Butterball employee stomped on a bird's head until her skull exploded, another swung a turkey against a metal handrail so hard that her spine popped out, and another was seen inserting his finger into a turkey's cloaca (vagina)." At ButterballCruelty.com you can watch video from the investigation and hear a Butterball worker boasting, "I kicked the fuck out of the motherfucker." Tough language, tough guy, considering you are talking about a defenseless twenty-pound animal.

Raeford Rocky Rehearsals

Lest you think the PETA investigators found abhorrent aberrations, Cleveland Ohio's Fox affiliate has broadcast video from an undercover investigation done by Mercy for

Animals at the House of Raeford Farms poultry plant.[10] The video includes behavior from workers similar to that at Pilgrim's Pride and Butterball, though with some special talents on show. We see one worker throwing punches at turkeys suspended on the conveyor belt as they moved past him on their way to the stun bath. We learn that he is an amateur boxer and thinks the birds make great punching bags for practice—live, conscious, and responsive, so even better than Rocky's hanging cow carcasses.

Kosher Kindness?

The Humane Slaughter Act dictates that animals should be unconscious before the fatal blow is delivered. As noted above, all birds are exempt. Also exempt are animals killed under religious Jewish and Muslim law. Under those laws, each animal must be killed by a swift cut to the neck—the idea being to protect the animals from unnecessary suffering. Also, the animal must be conscious before the cut. That law originally protected the animals by ensuring that the standards of treatment kept them healthy enough to walk to slaughter. It was also meant to protect humans by guaranteeing that animals so sick as to have died from transmissible diseases could not enter the human food chain.

Sadly, people controlling ritual slaughter have chosen to stick to the letter rather than the spirit of the law. The insistence on consciousness now means that religiously slaughtered animals suffer even more than most. Though humane slaughter laws make it illegal to shackle animals and lift them by their legs, we see the unfathomable exception for birds also extended to kosher slaughter and halal slaughter.

A *New York Times* story on a halal slaughterhouse reported:

> One by one, the largest animals are led into an enclosed metal pen where a chain is looped around one leg and the animal winds up hanging upside down.[11]

One wonders if *New York Times* readers would have considered (as we did a few pages ago), what it must feel like for a huge animal to be hoisted up by one leg.

In 2005 and 2006, the Department of Agriculture investigated violations of animal cruelty laws at the United States' largest glatt kosher slaughterhouse—the only one allowed to export to Israel—AgriProcessors, Inc. The *New York Times*, reporting on the findings, described undercover footage shot at the esteemed kosher slaughterhouse:

> It showed that after steers were cut by a ritual slaughterer, other workers pulled out the animals' tracheas with a hook to speed bleeding. In the tape, animals were shown staggering around the killing pen with their windpipes dangling out, slamming their heads against walls and soundlessly trying to bellow. One animal took three minutes to stop moving.[12]

That footage is online at www.goveg.com/feat/agriprocessors/.

Avoiding meat labeled kosher is no safeguard against buying meat from animals who were not stunned before slaughter—and that isn't just because of the 5 percent error allowance or because enforcement of humane slaughter laws is generally lax. Only the forequarters are used for kosher meat, so the sirloin and T-bone steaks from steers killed in the kosher manner are sold as nonkosher meat.

Highway to Hell

Many people say they are comfortable eating meat if they feel the animals are given good lives followed by "one bad day." It is probably fair to say that being hung upside down and bled to death goes beyond anyone's concept of a "bad day." Plus, the horrific end is usually dragged out over many days.

Factory-farmed animals see their first glimpse of natural daylight as they are loaded onto trucks that will take them to slaughter. The loading is horrific—cows and pigs are kindly encouraged with electric prods. The animals, many weak, sickly, and lame from spending their lives standing in small cages, are packed in together for shockingly long journeys on which they receive no food, water, or protection from the elements.

"Most of our fans go home where they have their dogs and cats who are part of their families. But then they go and have bacon and hamburgers, and they don't make the connection—this one I pet, and I eat this one—who has had the most brutal life imaginable on a factory farm" (**John Feldmann**, lead singer of Goldfinger, and producer of Good Charlotte and the Used).

Behind the Mask (2006), documentary produced and directed by Shannon Keith.

Courtesy of John Feldmann

"I get involved with animal rights because I love animals. And I really think that nine out of ten people, if they really knew what was going on . . . if they knew how fur was made, if they knew how animals were killed, if they knew about the testing on cosmetics, they wouldn't be down with it.

"That's because I think most people are good—and most people love animals. How can you not? You walk into your house every day and you see your dog or your cat and it makes you smile and it makes you happy. Well, the other animals out there are just the same.

"When I realized what was going on I said, 'I don't want to contribute to that. I want to be part of the good that's going on in the world.'

"That's also part of punk rock—taking a stand and making a difference in the world. And whether you are punk or you're not, you can make a difference.

"If you are thinking about getting involved in animal rights or being a vegetarian, it may seem like a hard thing. But I have a proud feeling every time I am at a restaurant and I say, 'No, I don't eat meat.' Just standing out and taking a stand in the world on something you believe in—it's a pretty good feeling" (**Benji Madden**, Good Charlotte).

Benji Madden interview available online at www.peta2.com/OUTTHERE/o-gcbenji.asp.

Poultry in Motion

Chickens raised for meat are called "broilers." They are raised in huge barns that hold tens of thousands of birds, are rounded up by catchers who are paid by the number caught—generally a couple of dollars per thousand chickens. A *60 Minutes* report[13] on the plight of chicken-industry workers showed those catchers each swinging ten chickens by the legs—five birds in each hand. The men literally pitched the chickens, like baseballs, into the trucks that would take them to the slaughter plant. When they missed the doors, the chickens hit the sides and bounced off—more like rebounding basketballs. But rebounding balls don't break limbs when they land—chickens do.

Long Day's Journey into Night

Animals are trucked from all over the United States, and even from Mexico and Canada, to slaughterhouses mostly in the Midwest. The *Des Moines Register* reported that about sixteen million pigs and more than a million cows are shipped into Iowa every year. More than a million of the pigs come from North Carolina, about a thousand miles away, and more than a million come from Canada. Calves are also shipped interstate. According to the president of the Iowa Cattlemen's Association, trucks carrying calves avoid stopping so the animals don't lie down.[14]

The USDA has announced plans to apply to trucking a nineteenth-century railroad law asserting that animals must be unloaded every twenty-eight hours. So what about twenty-seven hours of standing crammed together with no food or water, and no protection from the blazing heat in summer or icy winds of winter? No problem.

The special sounds good, but can I substitute the pork chop for a fried chunk of your left buttock?

And, surprise, even the twenty-eight-hour rule does not apply to turkeys and chickens. The National Chicken Council's Animal Welfare guidelines consider it acceptable if no more than 0.6 percent of "broiler chickens" die on the trucks. That translates to as many as fifty-four million birds every year.[15]

Toronto's *Globe and Mail* has reported that more than two million farm animals per year arrive dead at federally inspected slaughterhouses in Canada.[16] In the United States, estimates range from .1 percent to 1 percent of the ten billion or so land animals transported to slaughter every year.[17] That's between one and ten million, which is considerably lower than the National Chicken Council's number but still unconscionably high. *Feedstuffs* magazine tells farmers, "The average rate of loss of pigs during the trip to market [is] about 1 percent."[18] In summer they die from heat exhaustion. In winter they freeze to death. Perhaps those who arrive dead could be considered lucky; slaughterhouse workers have reported ripping skin as they unloaded live pigs who had frozen to trucks.

Animals who arrive dead, and therefore cannot be slaughtered under USDA standards for human consumption, are thrown on "dead piles." They may be thrown out as trash or may be salvaged for their leather. Or they may be rendered into protein paste for pet food or farm animal feed. Farm Sanctuary founders Gene Bauer and Lorri Bauston (Lorri now runs Animal Acres) tell the story of their first sanctuary resident, a sheep who they found while taking photos of a dead pile at a slaughterhouse. As the camera clicked, one of the sheep, buried in the midst of the carcasses, lifted her head. Gene and Lorri raced her to a veterinarian in the hope of euthanizing her humanely. But she wasn't sick—she had collapsed from exhaustion. Once nursed back to perfect health, and named Hilda, she became Farm Sanctuary's first ambassador.

Talk About a Downer

For years, animal advocates have been pushing legislative bills that would ban the slaughter of downed animals. A downed animal is one so sick she cannot walk into the slaughterhouse. She is usually conscious but cannot stand up—often because of a broken limb—so she is dragged in agony along the ground with chains, or pushed with a tractor, to her death. Because only animals slaughtered under USDA regulations can be sold for human consumption, which is the most lucrative sale, the plant owners avoid humanely euthanizing animals who collapse.

A short film by Mark von Schlemmer, which you can view on YouTube, includes undercover footage of a pig, trying to stand but unable, being beaten and beaten, then finally being dragged to slaughter along the ground by one ear.[19]

The agricultural lobby defeated a downed animal bill in 2003, just months before a dairy cow with mad cow disease entered the American food supply. While there is some question about whether that cow was downed, we know that downed animals are particularly likely to be diseased. The mercenary calculation that has led to slaughtering animals who cannot walk, rather than more humanely killing them where they lie, has put the public at risk.

Shortly after the announcement that a cow in Washington State was diagnosed with mad cow disease, the USDA implemented a policy prohibiting the use of downed cattle in human food. Though designed to protect humans, that law also removed the financial incentive for dragging cows to their death. In 2006, however, the new secretary of agriculture, Mike Johanns, announced that the department was considering allowing downed cattle back into the human food chain. Various animal protection groups have put forward the Downed Animal Protection Act in the hope of permanently banning all downed animals from the human food supply, a move aimed at providing at least some modicum of protection to both people and animals.

Factory Farming

The last day is the best day for most animals raised for food. Slaughter provides relief from a life of nothing but pain and frustration. The majority of the world's animals raised for food, and the vast majority in the United States, now live in what the industry calls concentrated animal feeding operations, or CAFOs. Commonly known as factory farms, they are more like factories than farms.

The Animal Welfare Act of 1970 established standards for the confinement of animals, requiring that cages "provide sufficient space to allow each animal to make normal postural and social adjustments with adequate freedom of movement." It exempts animals used for food.

Intense confinement to the point where they can barely move is not all the animals suffer. State anticruelty laws generally exempt standard agricultural practice. Standard practice includes dehorning, debeaking, tail docking, and castration—all without anesthetic.

Life in a Coffin—The Sow Gestation Crate

As I noted in the opening chapter of this book, I will always remember my first reaction to a brochure put out by the Humane Farming Association. It showed a picture of a sow in a cage called a gestation crate. The crate was just slightly larger than her body, leaving her no way to turn around or even lie down with her legs comfortably outstretched. She was locked in place. The brochure explained that she would be kept most of her life in a crate like that, or in a similarly sized farrowing crate just after she gave birth to each litter. I looked at the photo with horror, then told myself that those animal rights loonies had found one farm somewhere in the world that kept pigs that way, and they were trying to tell us it was typical. A year later I read *Animal Liberation* and learned that it was typical. Today 80 to 90 percent of the millions of breeding sows in the United States live in those crates.

The National Pork Producers Council asserts that keeping sows in crates is treating them well.[20] But the public doesn't agree. When the matter has been taken to voters by ballot, as it was in Florida in 2002 and Arizona in 2006, the crates have been banned. (Next stop California in 2008. If you live there but have not registered to vote because you don't care much for politics—register now if you care for the animals.)

After the Arizona ballot initiative, Smithfield, the world's largest pork producer, saw the writing on the wall and announced that it will voluntarily phase out sow gestation crates elsewhere.[21] A *Wall Street Journal* article noted, however, "The crates at Smithfield's farms will be phased out completely by 2017. The company also contracts with farms. At those farms crates will have to be phased out by 2027." While that's a welcome step in the right direction, it is a step too slow. We cannot accept, for another decade or two, the keeping of large, intelligent, and sensitive animals in crates.

"Factory farming
is just disgusting.
All you have to do
is look at a couple
of photographs,
and it's not hard to
figure that out"
(**Chris Walla**,
Death Cab for Cutie).

Chris Walla interview available online at www.peta
2.com/outthere/o-DCFC.asp.

Communal Crates

Gestation crates are not the only horror suffered by pigs. Once a sow has given birth, her piglets are taken from her at a few weeks of age and moved to warehouses where they live crammed, communally, in barren indoor pens. Those pens might as well just be called larger crates given what they offer—nothing. The pigs' tails are cut off and the males are castrated without anesthetic.

An article in *Rolling Stone* reminds us that phasing out gestation crates hardly means a decent life for pigs on factory farms. Here is how reporter Jeff Tietz describes the conditions at a farm run by Smithfield:

> Smithfield's pigs live by the hundreds or thousands in warehouse-like barns, in rows of wall-to-wall pens. . . . Forty full-grown 250-pound male hogs often occupy a pen the size of a tiny apartment. They trample each other to death. There is no sunlight, straw, fresh air or earth. The floors are slatted to allow excrement to fall into a catchment pit under the pens. . . .
>
> The temperature inside hog houses is often hotter than ninety degrees. The air, saturated almost to the point of precipitation with gases from shit and chemicals, can be lethal to the pigs. Enormous exhaust fans run 24 hours a day. . . . If they break down for any length of time, pigs start dying. . . .
>
> Taken together, the immobility, poisonous air and terror of confinement badly damage the pigs' immune systems. They become susceptible to infection.[22]

The article tells us that millions of factory-farmed pigs—one study puts it at 10 percent—die before they make it to the killing floor. Many die on the farms, and, as noted above, many die at four to six months on their hideous journey to slaughter.

Chickens Can't Stand in Standing Room Only

"Broiler chickens" never feel the sun on their backs and never scratch in the dirt. They are crammed into huge sheds where bird droppings are cleaned out about once a year after many flocks. The birds are genetically engineered to grow fast—reaching full market weight in six weeks—and to have huge breasts. The huge-breasted birds grow so heavy that many cannot stand up. When not standing, they sit in excrement that burns painful blisters onto their breasts.

The unnaturally fast growth also kills many birds from heart attacks and causes lameness and immense pain. (While chicken breeders question whether lame birds are in pain, an experiment conducted in the 1990s showed that lame chickens are far more likely than normal chickens to choose feed spiked with anti-inflammatory medication.[23]) Sometimes chickens' legs break underneath their weight. But the advice from *Avian Advice*, an industry magazine, is that in order to be profitable, "it is better to get the weight and ignore the mortality."[24] Sounds like something you might hear on *America's Next Top Model*.

As they get large enough for market, the birds take up more room and are eventually packed in so tightly they can hardly move. Many die from the stress. The following quote is from the *Commercial Chicken Production Manual*:

"Limiting the floor space gives poorer results on a bird basis, yet the question has always been and continues to be: What is the least amount of floor space necessary per bird to produce the greatest return on investment?"[25]

A Hard Act to Swallow—Foie Gras

Two or three times per day long metal pipes are shoved down the throats of ducks on foie gras farms. Each little animal has over a pound of cornmeal pumped into his stomach every day. Aficionados say the practice takes advantage of the ducks' natural tendency to gorge themselves on grain and store fat for long migrations. But if they would naturally gorge themselves, why force-feed them? On MSNBC, Ron Reagan commented, "Last time I checked, there was no natural tendency on the part of ducks to shove stainless steel tubes down their throats and pump in huge amounts of half-cooked corn."[26]

Foie gras literally means fatty liver, but it is actually diseased liver that has expanded to five to ten times its normal size. Imagine how your gut would feel if your liver was ten times its normal size.

Many of the ducks die before they are slaughtered. In 2003, San Francisco ABC viewers saw undercover footage taken by the Animal Protection and Rescue League at the Sonoma Foie Gras farm. Here is how reporter Dan Noyes's voice-over described it:

> The activists found barrels of ducks that died before their livers could be harvested, others still barely alive. They also watched ducks too weak or overweight to defend themselves against the rats at Sonoma Foie Gras. Rats were eating these two ducks alive and you can see evidence of similar battles on several other ducks.[27]

Shot by the terrific activist Ariana Huemer, the Animal Protection and Rescue League video shows rats munching on the back ends of ducks who are no longer able to turn around and fight them off. You can see it on YouTube at www.youtube.com/watch?v=LmyhQXom8eU.

The Canadian Network CTV has aired video from an undercover investigation carried out by Global Action Network at a Quebec foie gras farm. It includes shots of tiny ducklings struggling toward the top of bins in which they are being smothered. The reporter explains, "All the females end up in the garbage where they just suffocate to death. It's because they produce smaller livers."[28]

The production and sale of foie gras has been banned in many countries, in the state of California (effective 2012), and in the city of Chicago. Animal advocates are pushing for more widespread bans.

"Thanking the Turkey Who Changed Thanksgiving"

The following audio commentary, by Karen Dawn, aired on Washington Post Radio *Thanksgiving Day, 2006*

At Thanksgiving, I remember Olivia. I met her in the year 2000 at Poplar Springs Farm Animal Sanctuary near Washington, D.C. Having fallen for Babe, the movie star, I thought I was there to meet the pigs. But Terry, the sanctuary owner, started our tour at the turkey coop.

Terry opened the gate, and introduced Olivia. Olivia had been living on a turkey factory farm till Hurricane Floyd wiped it out. Ironically that hurricane saved Olivia from a particularly gruesome death; turkeys are not covered under federal humane slaughter laws.

Olivia hobbled through the gate—"hobbled" because the ends of her toes had been cut off. That's standard practice on farms where jam-packed birds sometimes injure each other. For the same reason the end of her beak had been seared off. That procedure is painful, since turkeys use their beaks to explore for food, so they are loaded with nerve endings. But Terry explained that it is cheaper to hack off the ends of toes and beaks than to give animals enough space.

I was sitting cross-legged on the grassy hill near the coop. To my surprise, Olivia limped in my direction. She came close enough for me to reach out and touch her—gingerly. She moved farther in, and I could pet her. It was surprisingly like petting my dog.

I reached my fingers under the outer feathers on her back and could feel a layer of soft down underneath. I had only ever felt that down in luxury pillows. How odd, and lovely, to feel it warm, on a living being.

Within a couple of minutes, Olivia had edged herself into my lap! I continued to move my fingers through her down. She laid her head in the crook of my elbow. She fell asleep. I fell in love.

And I adopted her. I couldn't take her home, but every year her photograph is in the middle of our Thanksgiving table, surrounded by a vegetarian feast.

Olivia showed remarkable longevity for a modern turkey. Bred to be deformed with a grotesquely huge chest with lots of "white meat," she'd been too weak to stand when she had arrived at the sanctuary. But apparently a home with space to move, grass to enjoy, sunshine in which to bathe, and loving care had given her the will to live. And live she did, happily, until 2005 when we received the sad news that the sanctuary's lovely little turkey ambassador, my little ward, had died of cancer. We've since adopted other rescued turkeys and added more photos to our Thanksgiving table arrangement. But Olivia's will stay in the center. She changed our Thanksgivings forever, and nobody can ever take her place.

Washington Post Radio, November 23, 2006. You can listen to the piece online at ThankingtheMoneky.com/op-eds.htm.

This Steer's Life

Cattle raised for meat have a half-decent start in life. Whereas the male calves of dairy cows are carted off to veal crates, calves from cows raised for meat spend about six months with their mothers. It is usually they that we see grazing as we head along highways in farm country.

The calves do experience severe trauma at about two months of age, when they are branded and castrated without anesthetic. But otherwise the first few months of their lives are probably as most people imagine the lives of farm animals to be.

Then they go to feedlots.

"This Steer's Life" is the title of a *New York Times Magazine* cover story in which Michael Pollan traces the life of a steer from insemination to slaughter. This is how he describes his first sight of a feedlot:

> Cattle pens stretch to the horizon, each one home to 150 animals standing dully or lying around in a grayish mud that it eventually dawns on you isn't mud at all.[29]

(Later in this chapter, in "There's Shit in the Meat," I will look at the human health implications of eating animals who live in their own waste.)

Those cattle pens have no protection from the elements. The temperature on summer days in Texas often goes over one hundred degrees, yet an article in the *Journal of Animal Science* reported that "shade is generally not used in commercial feedlots in West Texas because it is not thought to be cost-effective."[30] Though feedlots lose animals to heat exhaustion, they do not lose enough to make it economically viable to provide shade. The industry can accept all of the animals being in horrendous discomfort, as long as most don't die.[31]

It used to take five years of grazing to get a steer to market size. Pollan writes that now six months of grazing and fourteen months in a feedlot will do the trick. That is due to "enormous quantities of corn, protein supplements—and drugs, including growth hormones."

As cows don't naturally eat corn, it disturbs their digestive systems so badly that it can kill them if not accompanied with antibiotics. The antibiotics stop most of the animals from dying, but do not prevent acidosis, which develops because corn makes their rumen unnaturally acidic. Acidosis feels like bad heartburn—with no Prilosec. Nor do the antibiotics assuage the pain of liver disease. More than 13 percent of feedlot cattle are found at slaughter to have abscessed livers. Moreover, antibiotics added to cattle feed have led to the evolution of new antibiotic-resistant "superbugs." (I will discuss that further in "Added Dangers," later in this chapter.)

Pollan suggests that eating only grass-fed beef is the ethical solution to the horror of feedlots. His article tells us, however, that he had witnessed every part of the steer's life until the final moments. That's because "the stunning, bleeding and evisceration process was off limits to a journalist." Pollan had heard reports of cattle waking up after stunning—only to be skinned alive. But he was comforted when he spoke to Temple Grandin, who assured him that things are much better than they used to be. As I discussed in "Dying Piece by Piece," however, even the improved standards allow for extreme suffering for 5 percent of animals—many thousands per day. If Pollan had, like Eric Schlosser, been on the slaughter floor and seen a slaughter gone wrong—if he had witnessed his steer's life all the way through death—one wonders if he still would be such an enthusiastic proponent of grass-fed beef from cattle killed in standard U.S. slaughterhouses.

Karen Dawn

kill a Horse to Eat a cow

Ranchers graze their cattle on the public lands that have been home to wild horses for centuries. The horses numbered two million in the 1900s but now number only in the tens of thousands. Cattle ranchers are subsidized, paying only nominal fees, sometimes one cent per acre, or one cent per cow. Since horses compete for grass, the ranchers contend that the horses hurt the land. A comprehensive study issued by the United States General Accounting Office, however, reported that "wild horse removals have not demonstrably improved range conditions," and that "domestic livestock consume 20 times more forage than wild horses." It also reported that "the behavior patterns of the horses made them less damaging than cattle in vulnerable range areas" and that in many areas "poorly managed domestic livestock grazing" is the primary cause of degradation.[32]

The cattle ranching industry has chipped away, for almost four decades, at a 1971 law aimed at protecting wild horses. For horse lovers, the removal of mustangs has to call into question the assertion that the ethical answer to the problem of the hideous treatment of cattle in feedlots is to eat meat from animals who have been allowed to spend their lives grazing. The grazing of cattle on public land in the United States is responsible for the decimation of the wild mustang population. (We will look at cattle grazing's impact on the environment, such as on South American rain forests, in chapter 7.)

A Mane Course

What happens to the mustangs? Close to 100,000 horses, including many thousands of wild mustangs, are killed in slaughterhouses every year. The Government Accounting Office report cited above noted that there are too many horses being removed for the adopt-a-horse program to absorb and that "most adopters sold thousands of wild horses to slaughterhouses." Thus the symbol of the American West becomes French gourmet dinner so that Americans can eat cheap beef steak. Since 2007, when existing U.S. slaughterhouses became inactive, unwanted horses have endured trips to Mexico to be slaughtered in plants where no humane laws apply. Congress is currently considering the American Horse Slaughter Prevention Act, which would save all horses, includ-

© Anthony Freda

ing mustangs, from slaughter in the United States and also from being transported to slaughter elsewhere. (See "Horsing Around with the Law" in chapter 8.) But even that act would not save the mustangs from the horror of the roundups.

I recently watched the classic 1961 film *The Misfits*—classic as it was the last film for both Clark Gable and Marilyn Monroe—and for some horses who never received glowing obituaries.[33] The story culminates with a mustang roundup—and brings home the horror. We see cowboys take off across the plains in a truck, chasing terrified horses who may never have seen a truck before. They drive up alongside a running animal and lasso him, then tie the rope to a spare tire, which they throw to the ground. The horse, lugging the tire, will soon become exhausted and stop running. The cowboys can therefore leave him while they chase and lasso other horses. After the chase, they backtrack and find the horses with the attached tires. They wrestle them to the ground and tie their legs together, so that the buyers can come get them the next day. Perhaps the most heartbreaking scene is the capture of a mare and her colt. As they run side-by-side, the mare is lassoed using the rope with an attached line designed to eventually anchor her in place. When the cowboys come back for her, the unroped colt is still with her. They don't even bother tying him up overnight as he isn't going anywhere without his mother. He licks her and nudges her with his hoof as she lies on the ground in ropes.

What we don't see in this film, thanks to its message of redemption, is the usual sorrowful next step—the corralling and trucking of these glorious creatures who had always lived free.

Despite its important message, I hesitate to recommend *The Misfits*—and I certainly wouldn't recommend it if those who made it were still making big profits from rentals—because horses do not act. When the horses in the film seem terrified, and frantic, it's because they are. Further, in the documentary *Making "The Misfits,"* the pilots employed for filming the chase scenes tell us that they were directed to fly lower and lower—until finally they felt a thump and realized they had hit one of the animals. That's show business. And that's why, in this book's chapter 3 section "My Dead Flicka," I recommend the beautiful and compelling animated film *Spirit*. That film, without animal actors, depicts both the glory of the wild mustangs and the tragedy of roundups—tragic whether they lead to death at slaughterhouses or lives of servitude.

Eating Nemo—Fish

In his classic work *Animal Liberation*, Peter Singer wrote, "Surely it is only because fish do not yelp or whimper that otherwise decent people can think it a pleasant way of spending an afternoon to sit by the water dangling a hook while previously caught fish die slowly beside them."[34]

They do not yelp or whimper but fish do make vibratory sounds, inaudible to our ears, indicating alarm and aggravation. They have a centrally organized nervous system; they feel pain. Their death, by suffocation when they reach the air, is long and drawn out. In the case of deep-sea fish, their death comes from decompression.

For years people questioned whether fish, with nervous systems different from ours, are able to feel pain. In 2003, however, behavioral biologists at Edinburgh University published results of an experiment in which they injected bee venom under the skin of some trout. The animals lost interest in food, their gills beat faster, and they rubbed the affected areas against the walls of their tank. When the fish were given painkillers, which would not remove the irritating substance but would alleviate the experience of pain, the fish acted normally.[35]

I liked the *MSNBC Countdown* host Keith Olbermann's take on those findings. In response to the director of a pro-angling group who said it was all supposition until we have proper, "bona fide evidence," Olbermann said, "Try jamming one of your fish hooks in your lip and see how bona fide that feels."[36]

Non Sequitur © 1993 Wiley Miller. Dist. by Universal Press Syndicate. All rights reserved.

"I have been vegetarian for about eight years. I think it is just a good, peaceful, moral thing to do. I remember a story that a friend of mine told me. They were pesco-vegetarians—they still ate fish. Then one day they fished for their fish, and they caught a fish that mated for life. And another fish swum around until it died because they had caught its mate. And now they are vegan" (Fall Out Boy's **Andy Hurley** [on the right]).

Andy Hurley interview available online at www.petatv.com/tvpopup/video.asp?video=fallout_boy&Player=wm&speed=_med.

Fish School

Though many of us have been taught that fish are practically automata who would never form attachments, some fish mate for life.[37] And fish communicate with sound. Scientists have noted different sounds that relate to different behaviors.[38] Further, popular myths notwithstanding, as I noted in the chapter 2 section "Confining Nemo," fish aren't stupid. A UK *Telegraph* article reported that tests on fish show they possess cognitive abilities outstripping those of some mammals and "that they can store memories for many months, confounding the belief that they forget everything after a few seconds." The article told us, "Australian crimson spotted rainbowfish, which learnt to escape from a net in their tank, remembered how they did it 11 months later. This is equivalent to a human recalling a lesson learnt 40 years ago."[39]

A man in Pittsburgh has trained pet goldfish to carry a football, shoot a soccer ball into a net, and even dance the limbo. You can check out www.fish-school.com to learn more about him and his fish. One of his fish learned those tricks a day after being brought home.[40] Gosh, it's been years, and I haven't been able to teach my boyfriend how to dance.

Gregarious Groupers

Divers tell stories of friendly and affectionate groupers who rub up against them. I have friends who dive who will eat steak but not grouper. This letter published in *Reader's Digest* captures a similar reaction to those animals:

> As an avid scuba diver, I got a kick out of your photo essay "Fish Face." It may be hard to believe, but fish really do have personalities. While scuba diving in the British West Indies, I met a fish who was such a character he had been given a name—Alexander—by a dive guide. The friendly grouper waved his fins as if to say, "Hello!" and swam with me as I explored. When I ventured out of his territory, he waited patiently until I returned. Alexander made a lasting impression on me—and on my diet. No more seafood buffets for me.[41]

Long Deaths on Long Lines

The family Sunday fishing trip is no fun for the fish who get hooked and then suffocate slowly. But compared to what fish go through when commercially caught, it is a blast. Commercially caught fish can take hours or days to die.

Hundreds of millions of fish are caught on long lines every year. Miles of lines, with baited hooks, are dragged behind boats. Once hooked, instead of being immediately dragged on board (to suffocate slowly), the fish are left alive, struggling at the end of the hooks until the trip is finished and the lines are pulled aboard—often sometime the next day.[42]

Gruesome Gill Nets

Other fish are caught in gill nets, which hang between floats or sometimes boats. The fish swim into them and are caught by the gills, so they are unable to swim out. Nets attached to floats will keep fish trapped and struggling for days, until boats come and haul them onto deck—where they flap around and suffocate. Many fish struggle so hard in the nets they injure themselves and bleed to death while still in the water.[43]

Gill nets don't kill only fish. After a beloved seal pup drowned in a net off Hawaii, a *Star Bulletin* editorial informed readers,

> The pup, dubbed Penelope, the first born on Oahu in eight years, is certainly not the only sea creature to have been harmed by the nets. The cheap monofilament webs trap everything that swims into them, including endangered species like turtles and monk seals. Weighted at the bottom, they also damage coral reefs, breaking off pieces or scraping them.[44]

A *Washington Post* story on endangered right whales, of which there are only 300 to 350 left in the ocean, told us that their biggest threats are being hit by ships or getting entangled in fishing gear.[45]

Trawling and "Trash Animals"

Trawling catches fish by dragging a huge net along the seabed. Animals—fish and others—who are caught in the trawl net are dragged along for hours. It is a popular method of catching shrimp. A *New York Times* article about shrimping quoted a scuba diving instructor who had seen the effect firsthand: "One day there's all kinds of fish, crab, octopus, maybe a turtle, and the next day it's empty, nothing but rocks and a sandy bottom."[46]

The *New York Times* article explained that dragnets haul up ten pounds of sealife for every pound of shrimp—"like gathering wild mushrooms with a bulldozer." The nontarget animals—fish, reptiles, and mammals—are called trash animals. Their bodies are thrown back in. While shrimping is the most famously wasteful, as only a tenth of the catch is desired and the rest is needlessly killed, there is astonishing waste in other catches. A story on the front page of Florida's *Herald Tribune* informed readers that masses of dead mullet can be found floating in the waters off the coast. Female mullet, valued for their roe, fetch five to ten times the price of male mullet, so fishermen dump the dead males in the water in order to make more room on their boats for females.[47]

Shockingly wasteful methods like trawling are depleting the waters of fish. In an article published in *Science*, ecologists projected "all commercial fish and seafood species will collapse by 2048."[48]

If they are right, people will do fine getting their protein from tofu, and their omega-3s from flaxseed and algae. But the dolphins, whales, seals, penguins, and pelicans will all starve to death and become extinct.

Dolphin Discards

It's hard to think of dolphins as "trash animals" inadvertently caught in the hunt for better meat. But the UK's *Independent*, reporting on Greenpeace efforts to interfere with two trawlers, told us:

> Campaigners say dolphins, who have to come up for air every six minutes or so, are dying in their hundreds, possibly thousands, each year, drowning entangled in the nets of these "pair" trawlers.

Their bodies are usually dumped back by the fishermen, their bellies slit to make them sink quickly. And by a combination of persuasion and harassment, the Greenpeace boats are trying to stop the fishing and save the dolphins.[49]

Feedlots of the Sea

When they were first created, fish farms were seen as the great solution to overfishing. A *Los Angeles Times* front-page story has since described them as "Feedlots of the Sea."[50] They are as bad for fish as factory farms are for land animals. Each farm holds up to a million fish. Instead of swimming for miles every day, exploring and hunting, salmon on farms are jammed into pens. They are dosed with antibiotics and pesticides to ward off disease and sea lice, and they are fed synthetic pigment to turn their flesh from gray to pink.

Their waste smothers the sea floor. The antibiotics have created resistant strains of disease that infect both wild and domesticated fish. The rampant sea lice infect wild salmon when they swim past the fish farms and when those on the farms escape, which is frequently. An independent biologist who collected more than seven hundred wild baby pink salmon around farms found that 78 percent were covered with a fatal load of sea lice.[51]

Above all, fish farms consume many more fish than they supply. Salmon are carnivores. It takes about 2.4 pounds of wild fish, ground into pellets, to produce one pound of farmed salmon. So, ironically, fish farms create the very depletion they were designed to resolve.

Animals other than fish die as a result of the depletion. A report on ABC's *Nightline* let us know that the emperor penguin colony made famous in *March of the Penguins* has lost 50 percent of its numbers over the past half century. Their main source of food, shrimplike creatures called krill, are being taken in enormous catches to be ground up and fed to farm-raised salmon.[52]

Sea lions who gather around salmon farms, hoping for freebies, are shot.

Gabi Endicott

Susan Delvin

"When I was a kid walking down the beach in B.C. there was always a mad clattering of little crabs ducking under the rocks. Today the beaches are quiet, barely a crab in sight. I can still kayak and find orcas jumping now and then but I wonder how long until even they will be a memory; when the chain will finally break in enough places that they're lost to history because we were unwilling to make adjustments in the way we live. They say 'those who adapt survive.' It seems to me that the longer we take to make changes the more likely we are to eat ourselves out of house and home. But, 'Hey, give me more of those sea bass, you only live once, let's see the shark fin soup' " (**Bruce Greenwood**, *Thirteen Days* and *John from Cincinnati*).

Fishy Business Practices

Attempting to avoid supporting factory fish farming, you could choose wild-caught salmon (who have struggled for hours on lines attached to boats), but you'll never know if that is what you are getting. The *New York Times* performed tests on salmon from eight New York City stores. The salmon was all marked "wild" and was selling for as much as $29 a pound, yet the paper found that the fish at six of the eight stores was farm-raised.[53]

Evan Agostini/Getty Images

Vegan Oscar-winning actress and now fabulous shoe designer **Natalie Portman** says, "I just really love animals and I act on my values."

"Quotes of the Day," *The Evening Standard* (London), April 25, 2006.

A Shark's Tale

There is shocking waste and cruelty in every stylish bowl of shark fin soup, which sells for up to $200 a serving. According to the *Financial Times*, the shark fin soup business kills a hundred million sharks a year and numerous species of shark are likely to become extinct in the next two decades.[54] The *New York Times* has reported, "Some sharks, like the hammerhead and the great white, have been reduced by upwards of 70 percent in the last 15 years, while others, like the silky white tip, have disappeared from the Caribbean."[55]

All sharks, including harmless whale sharks, are killed for their fins. Since shark meat is low in value and fishing boats are small, it is not economically viable for fishermen to keep the whole body. They slice off the fins and throw the shark back in while still alive. Unable to swim, he sinks slowly to the bottom of the sea.

why vegan—No Eggs or Dairy?

Some animal rights activists feel we simply do not have the right to take products produced by other species—that doing so exploits the animals. Others, with a less pure outlook, might find using animal products acceptable if the relationship is mutually beneficial—for example, a hen might choose to let us eat some of her eggs in return for food and shelter for herself and her chicks. Similarly, a cow might choose to share some of the milk she has produced for her calf. Unfortunately, we can't ask the animals, so we have to rely on the golden rule when determining what is ethical, asking ourselves what we would want in their situation. Regardless of whether people feel it could ever be ethical to consume animal products, the consumption of them as they are produced today contributes to cruelty many people would find unconscionable.

California's Unhappy Cows

The pretty picture of contented cows grazing in pastures is largely a picture of the past. On modern factory farms, dairy cows are made into milk production machines. They are reared on crowded dirt lots, or indoors, often in stalls in which there is not even enough room to turn around. Like humans, they must get pregnant before they lactate, so they are artificially inseminated, then kept lactating as long as possible after they give birth. Then they are impregnated again. To the dairy industry the male calves are waste products, or at best by-products that are carted off to veal crates, or to plants that make their stomach linings into rennet for cheese and render their bodies into protein paste for pet food or animal feed.

Though naturally the milk is produced for their calves, dairy cows are denied their young, who are taken away within a day or two. People who have worked on dairy farms say there is no more awful sound than cows bellowing for their babies—sometimes ramming themselves against their stalls. Temple Grandin, a world-renowned animal welfare expert who has been hired by major food companies, such as McDonald's, to supervise animal welfare improvements, has said, "When cows are weaned, both the cows and calves bellow for about twenty-four hours."[56] Michael Pollan (who is not an animal rights activist) wrote in the *New York Times*, "Weaning is perhaps the most traumatic time on a ranch for animals and ranchers alike; cows separated from their calves will mope and bellow for days. . . ."[57]

People are surprised to hear about such shows of emotion from an animal presumed to be simple. But perhaps that emotion is linked to her simplicity. After all, she can hardly distract herself from her loss by reading or writing books or watching television. After she gives birth she doesn't worry about her career and wonder how long she should stay away from it in order to care for her baby but still "have it all." To a cow, raising her calf is *it all*. When we deprive her of that one joy in order to satisfy the odd human craving for the milk of another species, is it really surprising that she bellows for days?

Cows have been known to do more than bellow and ram walls. I remember when I was growing up I saw a television report from England about a cow whose calf had been sold and trucked to another farm. The next morning the people who had bought the calf found him inside his stall, the gate of which had been knocked off, nursing from his mother. She had broken out of her own stall, wandered the countryside, and found her calf miles away. A similar story made the front page of a West Virginia newspaper in

2005. Winnie, a 1,050-pound Charolais, was sold without her calf, named Beauty. The next morning, Winnie's new owners saw that she had escaped through a hole in the fence. She was found twenty miles away, just across the river from her old home, where her calf had remained. Winnie had trekked through woods and people's backyards. The farmers recaptured Winnie by bringing out Beauty and leading the calf in the direction they wanted the mother to go. Winnie's new owners then decided to also take Beauty, and to keep the cow and calf together.[58]

A cow might also bellow from the pain of mastitis, an infection of the udders that causes swelling and can make milking excruciating. According to an article published in the *Journal of Dairy Science*, 30 to 50 percent of heifers suffer from it.[59] Mastitis is exacerbated by milking machines and by the hormones given to cows so that they produce more milk. It is treated with antibiotics that end up in that milk. Dairy farm video reports, which you can view at www.unhappycows.com, show cows with udders so swollen they are dragging on the ground. A *New York Times* article described mastitis-infected cows who "begin to fail from the stress of carrying an udder that can weigh as much as a full-grown man."[60]

Under natural conditions a cow would live to be about twenty. For a factory-farmed dairy cow the intense cycle of pregnancy and lactation continues for about four years, after which she is no longer producing at a profitable rate, so she is slaughtered for hamburger meat. At least 20 percent of U.S. hamburger meat[61]—more like 75 percent in the New England area[62]—comes from dairy cows. Bill Niman, the founder of Niman Ranch, says, "The

And remember, kids — not only is milk a fattening health hazard, but when *you're* drinking milk, it means a sad & lonely calf somewhere *isn't!*

Aa Bb Cc
2 + 2 = 4

BUZZKILL the LUNCHLADY

Bizarro © Dan Piraro/King Features Syndicate

public doesn't realize the fast food and whole hamburger business is totally dependent on the old dairy cows."[63] Thus the fate of the dairy cow—the horrendous journey to slaughter, and the horrifying day there—is exactly the same as the fate of the "beef cow." Only her life is different. It's even worse.

Babes in the Wood Crates

What would you do if your neighbor kept her puppy permanently caged, never letting him out to exercise or relieve himself? And what if that cage was a narrow crate and the puppy was chained by the neck so he could not turn around or even lie down with his legs outstretched? You'd probably call the police and have her charged with animal cruelty. That's how the vast majority of veal calves are treated. It's legal for precisely that reason—legal because it is done to the vast majority and is therefore considered standard treatment. Standard agricultural practices are exempt from the anticruelty statutes of most states.

Veal calves are the waste product of the dairy industry. A day or two after birth they are taken from their mothers and placed in crates. They are fed an iron-deficient diet to make them anemic so their flesh will stay the pretty light color so enjoyed by gourmands. Desperate to satisfy their craving for iron, calves will attempt to lick their own urine and feces—that's why they are chained in their stalls, unable to turn around. Crates must be wood. One veal-production magazine warned its readers, "Keep all iron out of reach of your calves."[64] Even iron nails cannot be used in stalls, as the calves will lick at them and ruin the light color of their flesh.

When veal calves are slaughtered at sixteen weeks they are often too sick to walk. One out of every ten veal calves dies in confinement.

Mutts © Patrick McDonnell/King Features Syndicate

The Los Angeles Times, August 13, 2005

"Got Milk? You've Got Problems"

Op-Ed by Karen Dawn

Dairy cows have overtaken automobiles as the No. 1 air polluter in parts of California, according to a *Los Angeles Times* article. A *New York Times* editorial discussed "the eye-stinging, nose-burning smell of cattle congestion in rural California," acknowledging that something had to be done. What nobody wants to say, in this land of milk and cookies, is that we shouldn't be drinking cow's milk.

In the last edition of his *Baby and Child Care* bible, Dr. Benjamin Spock made it clear that cow's milk is for baby cows, not for human children. He wrote that it was "too rich in the saturated fats that cause artery blockages" and that it "slows down iron absorption." He suggested that it may cause ear and/or respiratory problems, and may be linked to childhood onset diabetes. He stressed that infants should drink only human breast milk and older children should try soy and rice milk products.

But the dairy industry would rather you didn't know that. As it spends millions of dollars telling us that milk consump-tion will help us lose weight, it would rather we didn't see a study published in the June issue of the *Archives of Pediatrics & Adolescent Medicine*. The study found that children who drink more than three servings of milk daily are prone to becoming overweight, even if it is low-fat milk. Neither does the industry advertise the Harvard School of Public Health finding that 15% of whites, 70% of African Americans and 90% of Asians are lactose intolerant.

The dairy industry prefers to scare us with tales of brittle bones, hoping we don't notice studies showing that people in Asia, who consume almost no dairy products, have a significantly lower rate of hip fractures than people in "got milk?" America. Consistent with those results is Harvard University's 1997 Nurses Health Study, which followed 78,000 women over a 12-year period and found that those who consumed the most dairy foods broke the most bones.

And a study published just this month in the International Journal of Cancer found a 13% increase in ovarian cancer risk in women who increased their lactose intake in amounts equivalent to one glass of milk per day.

Men don't need milk either. A Harvard study published in 1998 linked high calcium consumption to prostate cancer, and in this week's news, we learned that Dean Ornish's low-fat, vegan diet (no dairy) may block the progression of that disease. While touting its products as a fundamental part of a healthy diet, the dairy industry won't rush to tell us that Scott Jurek, who just won the Western States 100-mile run—for the seventh time in a row—is vegan.

Now, we learn that the dairy industry may also be harming our children by polluting the air. The *Times* article quoted an attorney for the Center on Race, Poverty & the Environment, who said that in Fresno, in the center of the nation's dairy industry, one in six children carries an inhaler to school.

Instead of protecting us, the government aligns itself with the dairy lobby. The California Milk Advisory Board, a government agency, playfully took advantage of society's increasing concern for animal welfare with its phenomenally successful "happy cows" campaign, which shows extended bovine families grazing in meadows.

People for the Ethical Treatment of Animals sued the board for false advertising, arguing that most California dairy cows live miserable lives on overcrowded dirt lots. They are artificially inseminated annually, because they don't produce milk without pregnancies, and are pumped full of hormones so that they will give 10 times as much milk as they would naturally. Their calves are carted off to veal crates. Then at about age 5, the "happy" cows are turned into hamburgers. PETA's suit failed—on the grounds that government bodies are exempt from fair advertising laws. Government is free to say whatever it wants about the conditions in which cows live, or about the "health benefits" of milk.

Unfortunately, the government is unlikely to start running ads suggesting we follow Asia's lead and switch to tofu, or even collard greens, though both have more calcium per cup than cow's milk. But for your health, the environment, the animals, and for those kids in Fresno carrying inhalers, why not change your next Starbucks low-fat latte order to soy?

Photographs courtesy of Farm Sanctuary

"Can we, asked the philosopher, strictly speaking, say that the veal calf misses its mother? Does the veal calf have enough of a grasp of the significance of the mother-relation, does the veal calf have enough of a grasp of the meaning of maternal absence, does the veal calf, finally, know enough about missing to know that the feeling it has is the feeling of missing?

"A calf who has not mastered the concepts of presence and absence, of self and other—so goes the argument—cannot, strictly speaking, miss anything. In order to, strictly speaking, miss anything, it would first have to take a course in philosophy. What sort of philosophy is this? Throw it out, I say. What good do its piddling distinctions do?" (Elizabeth Costello in J. M. Coetzee's *Lives of Animals* [Princeton: Princeton University Press, 1999]).

"I believe that, both spiritually and physically, it is bad for you to consume animal products, and have found this transition to be natural and easy" (**Russell Simmons**).

Russell Simmons, *Life and Def: Sex, Drugs, Money, and God*, Three Rivers Press, 2002.

Egg-regious Cruelty

Egg-laying hens never nest and raise their young. They never flap their wings. They don't run or peck for food in the dirt. In fact they never feel earth or even floor beneath their feet.

Male chicks of laying hens are thrown out—they are smothered in big plastic garbage bags. Or they may be ground up live and made into feed for their commercially valuable sisters.

The females are debeaked with hot knives, which often cause blisters in their mouths. The knives cut through sensitive tissue similar to the quick of a human nail. As people tend to imagine that it is like clipping the end of a fingernail, I appreciated Natalie Jordi's description of debeaking, as she saw it in the film *Baraka*:[65]

> I still remember a particularly grisly shot of gloved hands holding a hot soldering iron to a chick's beak, its legs pedaling desperately under a bug-eyed, wild stare, a wisp of smoke curling up from its face. A professional de-beaker de-beaks twelve to fifteen birds a minute.[66]

Jordi continues:

> Which is why I was dismayed to learn that, contrary to what I wrote in my last post, organic chickens CAN be debeaked, and often are. They can also be force molted!

(I will cover the organic standards to which she refers later in this chapter.)

Debeaking is carried out to prevent the cannibalism that can occur under hideously crowded conditions. A typical cage is twelve by twenty inches, about the size of a single sheet of newspaper, and contains four to eight birds. The cages are stacked floor to ceiling in massive sheds.

They are made of wide-spaced wire so that waste will drop through. Hens living on the bottom tiers are showered with excrement from above.

The cage floors slope so that the eggs will roll aside. As birds' claws are designed for a life scratching in dirt, their life on sloping wire is life in pain. Damage to feet is the norm. In some cases toenails become permanently tangled in the wire, so the flesh grows around it, leaving the birds soldered to their cages. In those cases the birds starve, unable to move toward food or water troughs.

As of 2012 the European Union has agreed to ban the use of standard battery cages for egg-laying hens. No such legislative ban seems imminent in the United States, though, via ballot initiative, Californians may ban battery cages in 2008. Under pressure from animal protection groups, some companies are already choosing to use cage-free eggs, but 98 percent of U.S. eggs still come from battery-caged hens.

Bizarro © Dan Piraro/King Features Syndicate

When egg production declines, the hens may be shocked into molting, because the birds will lay a few more eggs after a molt. "Forced molting" involves withholding food, and often water, for a period of up to fourteen days.

Chicken Run Lessons

As the movie *Chicken Run* made clear, when their egg-laying days are over, egg-laying hens become chicken soup or pot pie. That goes even for those who live on the world's very best free-range farms, like the farm portrayed in the movie. "Organic" or free-range layers become organic chicken soup or organic pot pie.

As noted above, in the section of this chapter headed "Fowl Outs," hens are exempt from the Humane Slaughter Act. If they make it to the slaughterhouse, they are killed with no legal protection—they are shackled, electrocuted, and too often boiled alive.

DOCTOR FUN

FARLEY

© David Farley

"Menopause is easy - after you stop laying eggs, they eat you."

Zombie Chickens

The meat left on the bones of emaciated spent hens is generally not considered worth the cost of sending them to market. With no legal protection, the birds are disposed of by the cheapest means possible. They may be buried alive in huge ditches. An article from the *Press Democrat* in Sonoma County, the heart of California's chicken country, explains that the market for spent hen meat has collapsed, and farmers, who have to find ways of disposing their hens, have taken to gassing them and making piles of compost out of their bodies. Some hens survive the gassing, get buried alive, and suffocate in the pile of dirt and corpses. Others survive the gassing and burial. A concerned neighbor who had lived next to an egg farm is quoted: "We called them zombie chickens. Some of them crawled right up out of the ground. They'd get out and stagger around."[67] On death's door, those that survive the hideous ordeal die of starvation or are eaten by wildlife.

Scrambled Egg Layers

In 2003, a ranch owner in San Diego County disposed of thirty thousand nonproductive egg-laying hens *Fargo*-style, by feeding them into a wood chipper. Well, not quite *Fargo*-style—-the hens were dumped in alive rather than dead. They hit the blades conscious, some feet first, some breast-first. A San Diego deputy district attorney failed to prosecute the farmers, finding no criminal intent as they "were just following professional advice" from two veterinarians. One of the veterinarians was on the animal welfare committee of the American Veterinary Medical Association.[68] That's a good point to remember. People, most importantly legislators, often look to that organization for guidance on the humane treatment of animals, forgetting that the group does not represent animals but is simply a trade group for veterinarians.

Label Laws and Label Lies

Wholesome Organics?

Eating organic animal products is probably better for your health than eating standard products. Animals raised under certified organic conditions are not pumped full of hormones and antibiotics. But the products may not be as pure as you had hoped. The *New York Times* reported, "The Senate and House Republicans on the Agriculture appropriations subcommittee inserted a last-minute provision into the department's fiscal 2006 budget specifying that certain artificial ingredients could be used in organic food."[69]

People often mistakenly assume that an organic label means the animals have been treated well. It is true that in many cases their treatment is better than on standard factory farms. Organic standards do not allow the confinement of egg-laying hens in battery cages, or sows and calves in crates. Some of the animals may spend time on pasture. Yet still, economics comes first. Pigs love to root around in the dirt, but some organic farmers put rings through their pigs' noses to make rooting painful so the animals won't dig up the land. And many of the animals whose meat is labeled organic are housed largely indoors in big barren sheds.

The other common cruelties also remain: Animals are castrated, tails are docked, and beaks are clipped, all without anesthetic. As we noted earlier: egg-laying hens may be force-molted via starvation.

There are no special organic slaughterhouses. Like all other birds, organically farmed birds sent to slaughterhouses are killed without legal protection from egregious cruelty. Organically raised pigs and cows may suffer the same fate in U.S. under-regulated slaughterhouses as those raised on factory farms; they may "die piece by piece" or may be boiled alive.

A *New York Times* article explained that the popularity of wild boar has been growing in the United States with the demand for organic food. It described the capture of a hog as follows:

> Four of his short-haired scent hounds, which had been released earlier, began to bark from the darkness. Mr. Richardson jumped out of the truck and freed a black pit bull from a cage on the truck's flatbed. He chased after his pit bull into the darkness toward the barking hounds. . . .
>
> When he got to the baying dogs, the light on his miner's hat revealed that the pit bull, trained for just this purpose, had clamped onto the face of a feral hog.
>
> As he had done thousands of times before, Mr. Richardson, 58, pounced on the snorting beast and tied its feet together, immobilizing it. Within minutes, he had loaded the animal barehanded into a cage.[70]

The story includes a close-up photo of Richardson loading a terrified and struggling hog into his truck. When Richardson collects enough hogs, he sends the truckload off to slaughter.

One cannot deny that the boars Richardson sells have had better lives than most of the animals sold for food in the United States. From an animal welfare standpoint, eating those animals is preferable to eating factory-farmed pigs. But reading the violent story of their capture, and thinking about what they face at the slaughterhouse, can we pretend that this organic food product is humane?

Organic Dairy (Factory) Farms

A *Chicago Tribune* front-page story has called attention to the surprising treatment of organic dairy cows at Horizon farms, telling us:

> On a recent Wednesday morning, with crisp blue skies and temperatures in the low 80s, there was something missing from Horizon's pastures. Namely, there were no cows.
>
> Critics of Horizon, including two former workers, say the empty pastures are emblematic. The dairy's new management, installed a year ago, has been so obsessed with increasing production to meet the soaring demand for organic milk that it has mostly kept the cows in the barn, the former workers allege, despite a U.S. Department of Agriculture requirement that organic cows have access to pasture.[71]

The article lets us know how vague that requirement is. It explains, "Because of complaints that several large farms were exploiting loopholes in the regulations, the Organic Standards Board sought to clarify the guidelines last year. Under the new guidelines, organic dairy cows would be required to get about one-third of their diet from pasture four months out of the year."

To date, the USDA has not adopted the recommendations—regulations insisting that just one-third of a third, which equals one-ninth, of an organic dairy cow's diet must come from grazing. How many people would have thought organic dairy farms would have a problem meeting those minimal regulations?

Organic dairy cow babies become organic veal. Their mothers, when no longer highly productive, become organic hamburgers.

Welfare Wiggle Room

Animal welfare is becoming big business—so much so that meat labels denoting welfare standards were the subject of a front-page *New York Times* story.[72] The article notes the growing popularity of meat considered to be humanely raised, but it shares concerns over misrepresentations: some animal welfare labels allow questionable farming practices, such as cutting the tails off pigs and allowing animals to be raised entirely indoors.

© Mike Baldwin / Cornered

www.cartoonstock.com

The *Times* article includes a sidebar informing readers that "Certified Humane" and "Free Farmed" labels allow the castration of sheep, electric prods on cattle, and tail docking of pigs. "Animal Compassionate" standards allow electric prods but forbid castration and tail docking, and Animal Welfare Institute (AWI) standards, the best of the bunch, forbid all three.

The article entirely ignores the fact that all of the programs allow the castration of pigs, because pigs without balls taste better—not so gamey—unless the pigs are slaughtered young. But slaughtering pigs young, at low body weight, is not economically viable. Neither is the use of anesthetic for castration deemed economically viable.

Also, with regard to the AWI standards, the article tells us:

> The Animal Welfare Institute and "free farmed" allow nose rings for pigs; the rings make rooting more difficult and prevent the pigs from tearing up the ground.

As noted in the section on organic meat, pigs love to root, and those rings are painful. So while pigs raised under Animal Welfare Institute standards certainly do better than pigs kept in crates, the article makes it clear that they do not have entirely natural lives.

Most animals reared for AWI labels will go to standard slaughterhouses.

The Chicken and the Egg Labels

CNN covered the myth of free-range chickens. The reporter, Sean Callebs, explained that according to the USDA, free-range producers must provide birds access to the outdoors, but the access can be through one small door in a crowded coop. He said, "The industry admits most free-range chickens don't stretch their legs." He also explained that free-range chickens, like other chickens, are bred to grow huge chests. Therefore many live lame and in pain, while some birds with "tiny legs overburdened by bodies" die because they can't reach the water source.[73]

Cage-Free Eggs—Not All They're Cracked Up to Be

Occasionally I have bought eggs for my dogs from a little organic farm in Southern California. The climate is temperate. The hens spend their days in a large shaded coop, with plenty of nesting boxes. In the evenings, I have seen the hens wandering around the farm and eating their dinner from a huge heap of grain and vegetable scraps. They look happy. When the hens stop laying, they live with their flock until they die of natural causes. They are not slaughtered.

Courtesy of PETA

At a glance, this farm appears to be a rare place in which the human consumption of animal products is truly symbiotic and mutually beneficial. But what happened to the males? We know that on standard egg farms they are ground up after hatching. Unless the breeders who supply hens to idyllic-looking hen farms retire their roosters to some sanctuary of which I am not aware, the cruelty-free egg production story seems like a tall tale.

Moreover, please don't mistake this place, which produces a few dozen eggs per day, with the organic or free-range farms selling eggs to your supermarket.

Cage-free eggs, and those with other humane-sounding labels, aren't all they're cracked up to be. While eggs that are "Certified Organic," free-range, free-roaming, and "Certified Humane" come from hens who are much better off than those kept in battery cages, the labels don't signify hens pecking happily in a farmyard. Eggs sold under any of the above labels generally come from hens packed into huge sheds.[74] Just as with "broiler chickens," even when outdoor access is required (as it is for "Certified Organic"), it may be so little, and available through such a small opening in the huge shed, that few of the hens ever get outdoors.

All of the labels named above allow for painful unanesthetized debeaking, and all except for "Certified Humane" allow starvation to induce forced molting. "Certified

Humane" guarantees the most—the animals must be able to perform natural behaviors, such as nesting, perching, and dust bathing. But they need have no outdoor access at all—their lives are nothing like what we imagine life should be for hens.

Scamming Caring Consumers

For years United Egg Producers stamped "Animal Care Certified" on all egg cartons. What the label meant was that the executives had heard that their customers were concerned about animal welfare—so they stamped something welfare-sounding on the cartons. The eggs came from birds living in battery cages who were debeaked and force-molted via starvation. A mighty little grassroots group called Compassion Over Killing took video of hens at an "Animal Care Certified" facility. The *Washington Post* described it as follows:

> The videos—shot by Takoma Park animal advocates who say they have spent years sneaking into local poultry farms—show hens closely packed in wire "battery cages," some missing most of their feathers, with open sores and burned beaks, and dead fowl caged with the living.[75]

The activists took the matter to the Better Business Bureau, which deemed the "Animal Care Certified" labels misleading and in 2004 referred the matter to the Federal Trade Commission (FTC). The FTC ruled that the label must be removed from all United Egg Producers cartons by March 31, 2006. Now eggs certified by United Egg Producers carry a "United Egg Producers Certified" label. That should give you the warm fuzzy feeling that comes with a guarantee that the hens have lived crammed into battery cages, hardly able to move, let alone spread their wings, and have been debeaked without anesthetic.

The Swine Welfare Assurance Program, or "SWAP," label is on almost every piece of pork you find in the grocery store. It means that the pigs were raised on factory farms and subjected to tail docking and castration or hideous sow gestation crates—and that pork producers have also learned that consumers worry about animal welfare, so they too figured they should stick on a label that includes the word "welfare."

Bee vomit

Our taste for bee regurgitations is bad news for bees, about a billion of whom have traditionally been exterminated every year by the honey industry.[76] At the end of the season it is not worthwhile for the farmers to winterize the hives—it is often considered easier to burn them and start again next season. That's how we thank the honey-makers.

Many people who care about animals don't worry too much about insects. They may have been taught that insects don't have brains and are basically automata without sentience. But in the chapter 4 section "Silkworm Union?" we looked at personality types that scientists have discovered in insects. In his informative book *Drawing the Line*, Steve Wise discusses what we know about the sentience of various species and how that information should influence arguments for their rights. I will share, below, some information from his chapter on bees, which is an eye-opener.[77]

Breakfast at the Honeypot Ant Diner

Wise begins by describing an experiment in which bees flew into a decision chamber with either a blue or yellow sign at the door. If they exited by a tunnel marked with the same color as the entry door, they received a sip of sucrose at the end of the tunnel. Bees learned that task easily. They could also learn to choose the sign that did not match. And they exhibited what scientists call "cross model associative recall": They could be trained to exit yellow if they had smelled lemon and exit blue if they had smelled mango. Their learning was not hard-wired. They could learn to reverse the associations.

Bees communicate with intricate "waggle dances" by which they tell each other where to find nectar. They can send ten thousand distinct messages, detailing the distance, direction, and quantity of food.[78] Wise describes a study by the Princeton biologist James Gould in which Gould brought bees to a boat loaded with nectar that he had anchored in the middle of a lake. Once the bees had gorged, he let them fly back to their hive. The bees did the waggle dance telling the others that there was a huge supply of nectar in the middle of the lake—a seemingly impossible location. Just as you might ignore an ad for a fabulous $100,000 condo for sale in the West Village of Manhattan, assuming it to be a misprint, the other bees did not race to the middle of the lake. When Gould repeated the experiment with the boat on the far shore of the lake—perhaps more like Greenpoint in Brooklyn—and the bees returned to their hive and did the dance communicating the location, the other bees took off across the lake immediately. The comparative psychologist Marc Hauser has remarked that this study indicates that bees not only have symbolic cognitive mental maps but also have a mental tool for skepticism.

Scout bees investigate possible home sites and then dance information about them to the group. Researchers have seen the following: Bees have advertised a spot they have found, watched other bees dance about other spots, gone to investigate the competing spots, and then returned home and changed their vote.

Bees do not have a cerebral cortex. But Wise tells us that Carl Jung, having read the studies that provide evidence of bee language (now a mountain of evidence), said that if that kind of behavior were seen in a human being, "we would certainly regard such behavior as a conscious and intelligent act and could hardly imagine how anyone could prove in a court of law that it had taken place unconsciously." He said we are "faced with the fact that the ganglionic system apparently achieves exactly the same result as our cerebral cortex."

Since human beings can lose a part of the brain generally responsible for certain functions and then learn to use another part instead, it wouldn't seem such a stretch to infer that bees have evolved the ability to reason without the part of the brain we had thought was needed for reasoning.

Interestingly, Wise writes that he doesn't think the evidence supports Jung, but he presents nothing that effectively refutes it. He does, however, write, "The physical, economic and psychological obstacles to dignity rights for honeybees are overwhelming." My impression is that Wise was surprised to learn about the level of consciousness of bees (he says as much in the chapter) and he knew that accepting it would jeopardize his argument that rights should be based on levels of consciousness. It would jeopardize it because the studies suggest that at least some insects are highly conscious—ruled by thought as much as instinct—and the world is not ready to give consideration to insects. So Wise rejected what appears to me to be the most reasonable conclusion. Perhaps some others will read the information and will consider rejecting honey.

There are good reasons based on human selfishness for rethinking the way we treat bees. A *New York Times* front page story has reported, "More than a quarter of the country's 2.4 million bee colonies have been lost—tens of billions of bees, according to an estimate from the Apiary Inspectors of America, a national group that tracks bee-keeping. So far, no one can say what is causing the bees to become disoriented and fail to return to their hives." The article also told us, "Honeybees are arguably the insects that are most important to the human food chain. They are the principal pollinators of hundreds of fruits, vegetables, flowers, and nuts."

The reporter cites all sorts of weird and wonderful theories to explain the potentially devastating disappearance of the bees: "People have blamed genetically modified crops, cellular phone towers and high-voltage transmission lines for the disappearances. Or was it a secret plot by Russia or Osama bin Laden to bring down American agriculture? Or, as some blogs have asserted, the Rapture of the bees, in which God recalled them to heaven? Researchers have heard it all." Bacteria, fungi, and pesticides are other possibilities put forward.[79] A CBS *60 Minutes* report has suggested that new pesticides called neonicotinoids, derivatives of nicotine, could be to blame. They cause insects to lose their sense of navigation and could well be affecting the bees, causing what scientists have termed "Colony Collapse Disorder."[80]

As I learned of the crisis, I remembered that bees have language—ten thousand distinct messages. Is it therefore so far-fetched to wonder if perhaps rumor got around bee world that returning to the hives meant being exterminated at the end of the season? I don't know. But what we all know now about bees, as we are faced with their loss, is that we should not be so cavalier with their lives. Would it be so hard to treat them with respect, furnishing them with bountiful food as thanks for some of the honey they produce, while leaving them plenty for themselves and also offering them safe housing over non-productive seasons? Not at all—but first we would have to acknowledge that they matter. I thought we were being forced by the crisis to make that acknowledgment until I heard the *60 Minutes* reporter say, "Scientists are patiently trying to re-create 'Colony Collapse Disorder' in healthy hives in order to try to determine what's triggering it." There must be a better way.

Doug Hall/PCRM

Happy Hunting

On a camping trip I met some hunters who said, "What we can't stand is when people criticize us for hunting, then go to the supermarket and buy their neat little packets of meat made from animals who have lived their whole lives indoors in cages. At least the animals we eat had a life. And we have a right to kill them to feed our families. It's the law of nature."

I have nothing to argue with there. Shooting a wild animal is incomparably kinder than buying one who lived on a factory farm. And though I would personally hate to do it, I could never say it is unethical to kill an animal, if one must, in order to give one's family a nourishing meal.

In case you think I have just given out free passes for a hunt this weekend, I had better add that I would be shocked to find those parameters apply to anybody reading this book. In his engrossing book *The Omnivore's Dilemma*, Michael Pollan does a superb job of portraying factory farming, a disgraceful system he chooses not to support. He prefers to seek out grass-fed beef, or hunt for his own meat. Unfortunately, he writes that while he envies the moral clarity of the vegetarian, he pities him too. He writes, "Dreams of innocence are just that; they usually depend on a denial of reality that can be its own form of hubris. Ortega suggested that there is an immorality in failing to look clearly at reality."[81]

Pollan's reality, however, is that he lives in Berkeley, probably the easiest place in America to get a great vegetarian meal. Bless him for drawing his line a little higher than those willing to eat factory-farmed animals, but the line he draws puts his desire to taste meat above the fundamental and profound desire for life shared by the thousands of animals he will consume during his life. He lives in a city surrounded by those who draw the line higher than he does. I hope they smiled at his accusation that they are the ones immorally avoiding reality.

Ironically, Pollan's argument about vegetarians and reality comes at the end of a chapter in which he has described his first hunt, and done so with a glaring avoidance of reality. He and his friend Angelo had each taken a shot at a pack of pigs. Pollan writes, "One pig was down; another seemed to stagger." Then he writes:

I collected myself just enough to pump and fire one more before the pigs dispersed, most of them tumbling down the steep embankment to our left.

We ran forward to the downed animal, a very large grayish sow beached on her side across the dirt road; a glossy marble of blood bubbled directly beneath her ear. The pig thrashed briefly, trying to lift her head, then gave it up. Death was quickly overtaking her, and I was grateful she wouldn't need a second shot.

What of the pig who had "seemed to stagger" but had tumbled down the embankment with the rest? Pollan writes that he tried to go after that second animal but the embankment was too steep.

In the section "Hunting—The Great Divide," in chapter 7 of this book, I discuss in more depth the agony of animals who are injured by hunters but not killed outright. Even when an experienced hunter brings down an animal in a clean shot, there are ethical issues—the taking of a life that mattered to the animal, and that may have mattered to other animals, such as any young left motherless. But not all animals are brought down by hunters with single clean shots. Hunters kill and bring home tens of millions of animals every year and they hit millions of other animals who get away and die slowly in agony.[82]

Hunting is still less cruel than eating the products of factory farming, transport to slaughter, and under-regulated slaughterhouses. But it causes much suffering. Each of us can look at that reality, and then decide whether or not we wish to support it.

In Vitro Meat—The Best Thing Since Sliced Bread?

Scientists are working on lab-grown meat. According to the *New York Times*:

The process works by taking stem cells from a biopsy of a live animal (or a piece of flesh from a slaughtered animal) and putting them in a three-dimensional growth medium—a sort of scaffolding made of proteins. Bathed in a nutritional mix of glucose, amino acids and minerals, the stem cells multiply and differentiate into muscle cells, which eventually form muscle fibers. Those fibers are then harvested for a minced-meat product.[83]

Other articles report that scientists have grown something similar to fish sticks by slicing a bit of muscle from the abdomen of a goldfish and placing it in a serum solution to grow.[84] They used fetal calf serum, but maitake mushroom solution also works. The researchers battered the fish sticks, fried them in olive oil, and then subjected them to a sniff panel; the panelists weren't allowed to eat the fish sticks but said they smelled appetizing.

If it became common, lab-grown meat could have profound environmental consequences. In the United States alone, animals on factory farms produce 1.4 billion tons of waste every year. Laboratory meat production would not foul rivers with manure or heat up the atmosphere with methane. (See chapter 7 for more on that.)

The in vitro meat researcher Jason Matheny notes that lab-grown meat wouldn't harbor diseases, such as avian flu, or contribute to antibiotic resistance. And it could be tweaked to make it better for human health—for example, saturated beef fat could be swapped for healthy salmon fat.[85] Imagine McDonald's burgers that were relatively good for you and that didn't hurt the animals!

Indeed, from the standpoint of the animals, the repercussions could be revolutionary. Lab-grown meat could signal the end of factory farms. It could even be the end of slaughter. Matheny has said, "With a single cell, you could theoretically produce the world's annual meat supply."[86]

Many of the articles on the issue suggest that people will find lab-grown meat too hard to swallow. They talk about "the yuck factor." But yuck factors are surmountable. A lot of children feel pretty yucky when they first learn that spareribs come from Wilbur. But they get over it. I bet it turns out to be even easier for them to get over learning their meat came from a petri dish—after all, isn't that pretty much how it looks in the supermarket fridge section all wrapped up in plastic? As for adults, once you've seen *Fast Food Nation* and heard about the "shit in the meat" (you'll read more about it later in this chapter), in vitro meat, with no feces and no *E. coli*, might seem comparatively yuckless.

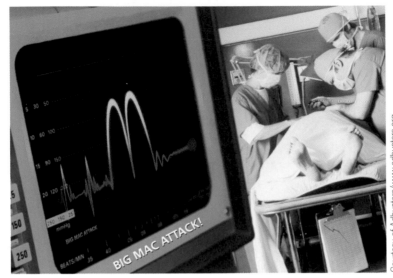

Human Health

Ronald McDonald Strikes Again

In the documentary *Supersize Me*, Morgan Spurlock followed the new standard American diet to the extreme. He ate nothing but McDonald's food three times a day for thirty days. His health deteriorated far more quickly than anybody would have imagined. His doctor advised him not to finish the experiment, as his liver was being damaged and was in danger of failing.

In April 2004, McDonald's chief executive, Jim Cantalupo, age sixty, died of a heart attack at a Florida fast-food convention. His successor, Charlie Bell, died of colon cancer nine months later, at the age of forty-four. Is it not reasonable to assume that executives take pride in liberally consuming their company's products? I guess the Happy Meals did them in. Some people might think the fries and shakes got them, but look at Dr. Robert Atkins:

Not long ago, millions of people embarked on the Atkins diet, which banished almost all carbohydrates and encouraged people to eat bacon and cheese omelets for breakfast, hamburgers without the bun for lunch, and steak with sides of chicken for dinner. Dr. Atkins was the poster child for his own diet, until 2004, when a year after his death at age seventy-two, his medical records were released. According to the medical examiner's notes, Atkins had heart disease—a myocardial infarction, hypertension, and congestive heart failure. He had earlier blamed a cardiac arrest on an infection, and had publicly boasted that he had no blockages. But according to his cardiologist, in 2001 his coronary arteries were 30 to 40 percent blocked. His wife confirmed in a statement that in the last three years of his life he had developed new blockage. He was also on heart-rhythm medication.[87] Are we sup-

posed to think that if he had not been on his own healthy diet, he would have had some more serious health problems? All of the above caused some skepticism with regard to the reports that Atkins's death had been caused by a slip on the ice outside his apartment—some speculated that a collapse on the ice was more likely.

Conversely, Donald Watson, the founder of the Vegan Society, recently died at the age of ninety-five. He had been interviewed at age ninety-two and was pleased to report that he had lived with hardly a day's illness.[88] We'll look more at the health benefits of vegan diets later in this chapter, under "Chew the Right Thing."

"There's Shit in the Meat"

In the film *Fast Food Nation*, an executive at a fictional fast-food chain explains to Greg Kinnear's character that there is a problem with fecal counts—in other words, "There's shit in the meat."[89] The film and Eric Schlosser's nonfiction book, on which the film is based, explain that due to the rate at which cattle are slaughtered in modern plants, workers cannot remove intestines with the necessary care, and often, at least daily, feces spray all over the rest of the meat. Leaking intestines aren't the only cause of fecal splatter: The animals spend their lives in stinking feedlots covered in their own waste—when they are cut up that waste goes everywhere.

In his book, Schlosser tells us that a single fast-food hamburger can contain meat from hundreds of different cows. And one animal infected with *E. coli* can contaminate thirty-two thousand pounds of ground beef. He cites a USDA nationwide study of ground beef taken at processing plants, which found that 78.6 percent of the ground beef contained microbes that are spread primarily by fecal material.[90]

Doug Hall/PCRM

Rather than clean up the feedlots and slaughterhouses, the meatpacking industry and USDA want to start irradiating meat. Irradiated meat is zapped with gamma rays or X-rays. The rays do not kill microorganisms, but instead change their DNA so that they cannot reproduce. The industry hopes to allay public concerns about eating anything irradiated by changing the name and calling the process "cold pasteurization." But the public would still eat shit—irradiated shit.

In 2006, America had a spinach scare. Spinach from an organic farm in California carried *E. coli*. It sickened many people and killed one elderly woman. An op-ed in the *New York Times* explained that the spinach farm wasn't the culprit. That particular strain of *E. coli* thrived in the unnaturally acidic stomachs of beef and dairy cattle fed on grain on industrial farms; up to 80 percent of dairy cattle carry it. Their infected manure contaminates the groundwater and spreads the bacteria to neighboring farms that grow vegetables.[91]

Dangerous Liaisons

It was fascinating to see the immediate recall of spinach, after one hundred people were sickened and one died, and to compare that with the reaction to tainted meat. Schlosser tells us that in the last decade, half a million Americans have been made ill by *E. coli* 0157:H7. Thousands have been hospitalized and hundreds have died.

Schlosser also tells us that in the 1980s "the USDA became largely indistinguishable from the industries it was meant to police. President Reagan's first secretary of Agriculture was in the hog business. His second was the president of the American Meat Institute (formerly known as the American Meat Packers Association). And his choice to run the USDA's Food Marketing and Inspection Service was a vice president of the National Cattlemen's Association. President Bush later appointed the president of the National Cattlemen's Association to the job."[92]

Fuzzy Recall

When public danger is so great that even the USDA would like to see a recall of contaminated meat, it cannot demand one. And a company that chooses to call one is under no legal obligation to inform the public, or even state health officials, that a recall is being conducted. Schlosser

"It's only the Ericksons, so why don't you just use the recalled hamburger meat."

tells us that Wendy's tried to recall about 250,000 pounds of beef, announcing "it had not been fully tested." Unfortunately, "the press release failed to mention that some ground beef from the same lot had indeed been tested—and had tested positive for *E. coli* 0157:H7."[93]

Everything Old Is New Again

Any doubt that our government puts the meat lobby before public welfare was surely allayed when the Food and Drug Administration (FDA) ruled, without any public hearings on the matter, that packaged meat could be treated with carbon monoxide. According to a *Washington Post* front-page story, carbon monoxide "gives meat a bright pink color that lasts weeks"[94] and "keeps even rotten meat looking red and fresh." Tuna may also be treated with the gas.[95] The products do not have to be labeled as treated.

The *Post* tells us that before its decision to allow the process, the FDA received a petition citing numerous studies reporting that consumers judge the freshness of meat, and decide whether or not to buy it, based on its color. But that didn't affect the ruling. One of the lobbyists for the gas technology explains that there is no need for concern (even though consumers have always relied on color judgments), because "when a product reaches the point of spoilage, there will be other signs that will be evidenced—for example, odor, slime formation, and a bulging package." Funny, those are exactly the same signs that told me it was time to dump my last boyfriend. The FDA ruling means it's just tough luck for consumers hoping for clues that their meat is going off before it is slimy. And really tough luck if they thought the FDA was going to protect them from the meat industry.

Doug Hall/PCRM

Karen Dawn

Method in the Mad Cow Madness

The USDA's double mandate—to keep meat safe and to promote the sale of American meat—also puts Americans at risk for mad cow disease. In 2003, the year that the first infected cow was found in the United States, of approximately thirty-five million slaughtered cows only twenty thousand had been tested. The *Seattle Times* imparted this astonishing information in February 2004:

> Nationwide, testing has plummeted after discovery of the mad-cow case near Yakima, announced Dec. 23. Only 1,608 animals were tested in January, down from 3,064 in December.[96]

A few months later, the *New York Times* editors expressed their lack of confidence that the nation was being adequately protected, noting that the USDA "keeps taking actions that suggest more concern with protecting the financial interests of the beef industry than with protecting public health." The piece shared this shocking tidbit:

> Just a few weeks ago, the department refused to let a small private company test its cattle for mad cow disease to satisfy Japanese customers.

The *Times* editors commented,

> That decision was incomprehensible, unless it was driven by a desire to protect the beef industry from pressure to conduct such tests on all 35 million cattle slaughtered annually in this country.[97]

How rampant is mad cow disease? We don't know. A study at Yale found that of forty-six patients clinically diagnosed with Alzheimer's, six were proven to have Creutzfeldt-Jakob disease (CJD) at autopsy.[98] Other studies have shown that mad cow prions can cause a disease with a molecular signature indistinguishable from sporadic

CJD. Therefore, there is no way to determine if the many deaths from CJD misdiagnosed as Alzheimer's are actually linked to mad cow disease.[99] So we cannot know how widespread mad cow disease is in the United States, or whether humans are infected. It seems that the government is in no rush to help us find out.

Added Dangers

The antibiotics fed to cattle (as described in the earlier section "This Steer's Life") lead to the growth of antibiotic-resistant superbugs, which end up in the people who consume them.

A *Washington Post* front-page story reported: "The government is on track to approve a new antibiotic to treat a pneumonia-like disease in cattle, despite warnings from health groups and a majority of the agency's own expert advisers that the decision will be dangerous for people. The drug, called

cefquinome, belongs to a class of highly potent antibiotics that are among medicine's last defenses against several serious human infections." The article tells us that "cefquinome's primary threat is that it may undermine the usefulness of the closely related human drug, cefepime" (brand name Maxipime). That drug "is the only effective treatment for serious infections in cancer patients and a reliable lifesaver against several other nearly invincible infections."[100] But apparently our lobby-fed legislators think that matters less than helping the beef industry keep the cows they sicken healthy enough to kill.

Added growth hormones, which help bring animals to market size quickly, are also suspected of having serious human health implications. That's something to think about when your kid sister starts shopping for her first bra in grade school.

Finger-Lickin' Bad for You

We see the same problems covered in the earlier section "There's Shit in the Meat" with chicken. "Broilers," jam-packed into huge sheds, are often so weak or disabled they are unable to stand, so they spend their days immersed in their waste, which covers the floor. At the slaughterhouse, some of that waste inevitably ends up in the meat.

Chickens, like other animals, are given antibiotics so that their living conditions don't kill them. An episode of the TV drama *Everwood*[101] portrayed a minister who broke out in hives every time he had sex with his wife. Everybody assumed it was psychosomatic. But it turned out she was on a diet that involved eating lots of chicken and he was allergic to the antibiotic given to the chickens. The story was funny but the point was serious—the animals are loaded up with antibiotics; so are people who eat them.

Don't, however, be tempted to think that downing those antibiotics protects you from the diseases carried by chickens. A 2007 *Consumer Reports* analysis of fresh whole broiler chickens nationwide revealed that 83 percent harbored campylobacter or salmonella. In case you thought things were improving, the article told us, "That's a stunning increase from 2003, when we reported finding that 49 percent

Bizarro © Dan Piraro/King Features

tested positive for one or both pathogens."[102] Most of the bacteria *Consumer Reports* tested showed resistance to one or more antibiotics. Premium brands were no safer than others.

Skinny Chick?

Chicken is not low in fat. Skinless roasted light meat is 18 percent fat, dark meat is 32 percent fat, and chicken cooked with skin may be 50 percent fat even once the skin is cut away, because the fat permeates the meat as it is cooked.

McDonald's McNuggets contain twice as much fat per ounce as a hamburger.[103]

More to Make You Chicken Out

When meat is cooked at high temperatures it forms heterocyclic aromatic amines, which are cancer-causing compounds. A report from the U.S. National Cancer Institute revealed that oven-broiled, pan-fried, or grilled chicken contains more of those carcinogens than red meat.[104]

The *New York Times* has reported that arsenic—yes, arsenic—is "deliberately being added to chicken in this country," as it "is used to kill parasites and to promote growth." The article also tells us that "arsenic is a recognized cancer-causing agent and many experts say that no level should be considered safe." The *Times* noted that some suppliers are starting to phase out the use of arsenic in chicken feed, but that in a study "none of the fast-food chicken purchased was arsenic-free."[105]

A *Minneapolis Star Tribune* article reported, "Nearly three-quarters of the chicken from conventional producers contained some arsenic, while a third of chickens from premium or organic producers contained arsenic. Every sample from fast-food restaurants had at least some arsenic."[106]

Mercury Rising in the Ocean

Fish is often touted for its health benefits, as it is high in protein and omega-3s. It is now also high in mercury. You might see the obvious upside here—we could soon dispense with thermometers and just look at each other for signs of climate change. But unfortunately mercury causes learning disabilities in children and neurological problems in adults. The FDA and the Environmental Protection Agency (EPA) have jointly warned pregnant women, nursing mothers, women of childbearing age, and young children not to eat shark, swordfish, king mackerel, and tilefish because of high mercury levels. The warning also cautioned those

© 2004, Mike Twohy. Dist. by The Washington Post Writers Group M2Ecomics@aol.com

"We're also pleased to offer a grilled mahi-mahi with an unseasonably low mercury level."

groups to consume no more than twelve ounces of fish per week, including no more than six ounces of canned albacore tuna.[107]

The *Chicago Tribune*, which did investigative pieces on the issue, also reported, "Testing by the *Tribune* showed that a variety of fish that consumers might assume are relatively safe actually contain high levels of mercury. For example, 15 of the orange roughy samples the *Tribune* bought had high levels." And it reported, "Some samples of grouper, tuna steak and canned tuna were so high in mercury that millions of American women would exceed the U.S. mercury exposure limit by eating just one 6-ounce meal in a week."[108]

Tribune investigators tested swordfish from Singapore for sale in Chicago. The results showed levels three times the legal limit.[109] That is especially shocking given that America's mercury limit is one of the highest in the Western world. Fish sold in America is allowed, for example, to have twice as much mercury as seafood sold in Canada.[110] Clearly our proud leaders know Americans are tough enough to handle twice as much mercury as our wimpy neighbors to the north.

A front-page *Wall Street Journal* story focused on a ten-year-old boy whose parents had been proud that their son always chose tuna over junk food. The boy ended up with mercury poisoning. The article told us that such groups as the National Food Processors Association, the National Fisheries Institute, and the U.S. Tuna Foundation had dissuaded the U.S. government from coming out with strict advisories regarding tuna consumption and that the American public had therefore been put at risk. It also told us that the limits set in the joint advisory finally issued in March 2004 by the FDA and EPA may exceed safe levels for some people, judging by a mercury risk assessment that the EPA produced on its own years earlier, before "food processors lobbied the administration."[111]

Doug Hall/PCRM

Indeed, when it comes to fish and mercury, we see again something similar to the double mandate we saw with the USDA. Marion Nestle explains that the FDA's regulatory mission "is to balance health risks against cost considerations, among them costs to industry."[112]

The *Chicago Tribune* commented: "In some cases, regulators have ignored the advice of their own scientists who concluded that mercury was far more dangerous than what consumers were being told. In other instances, regulators have made decisions that benefited the fishing industry at the expense of public health."[113]

Heart Disease Weekly has warned, "People who eat fish with a raised mercury content run a greater risk of coronary heart disease than previously thought." The article tells us, "Mercury, which is found in certain fish from environmentally contaminated areas, may counteract the health benefits of certain fatty acids that are also present in fish."[114]

In late 2007 the mass media widely reported the results of a study publicized by a "children's health coalition" announcing that the benefits of eating fish outweigh the risks. The *New York Times* report on the study, however, noted that the coalition's leadership took thousands of dollars from the fishing industry to promote the recommendations. And it noted that the public relations firm Burson-Marsteller, which publicized the findings, also represents the fisheries institute, and that the managing director of that public relations firm just so happens to be the vice chairman of the coalition touting the health benefits of fish! What's more, many member groups of the coalition were not warned of the impending publication of the findings, and have publicly disavowed them. The chairman of the nutrition committee of the American Academy of Pediatrics, a member of the coalition, went so far as to say, "We are appalled." The press release did have some of the desired effect: as a senior scientist with the EPA commented, the report "has created an artificial controversy." In truth, especially among child-bearing-age women, there is no controversy. Even the most pro-business government agencies warn women about the consumption of fish likely to contain mercury. [115]

Other Funky Fish Facts

Mercury is not the only toxin in fish. *Science* magazine has warned us about the danger of PCBs and other toxic chemicals in farmed salmon. It suggests that people should not eat it more than once a month.[116] Feel free to ignore that warning if your physician has recommended a diet high in toxic chemicals.

Recent tests done in a study by the University of Pittsburgh Cancer Institute's Center for Environmental Ecology have also shown that the fish in many streams, rivers, and lakes may be carrying enough chemicals that mimic the female hormone estrogen to cause breast cancer cells to grow.[117]

We know that human consumption of fish isn't good for fish. We can no longer just assume that it is good for us. I will discuss below, under "Taking Care with Vegan Fare," safer sources for nutrients found in fish.

Not Milk

I noted in my *Los Angeles Times* op-ed "Got Milk? You've Got Problems," reprinted in this chapter, that studies have linked cow's milk consumption to human cancer.[118] I also noted studies that have shown lactose intolerance to affect the vast majority of people of African and Asian descent.[119]

Hormones in cow's milk are another health concern. On YouTube you can watch a short fascinating documentary about two Fox 13 news correspondents in Florida, Jane Akre and the Emmy-winning reporter Steve Wilson, dubbed "the Investigators." The Investigators tried to publicize the use of bovine growth hormone to make cows produce more milk. They wanted to let viewers know the additive had been banned in Canada when scientists found that it was absorbed by humans and could have serious implications for human health—including possibly causing cancer. In the documentary, Akre explains that just before the story was set to air, Fox 13 heard from the agricultural company Monsanto's lawyers, and decided to pull the story and check it one more time. Wilson says, "But the bottom line was there were no factual errors in the story. Both sides had been heard from. Both sides had been given an opportunity to speak." Akre tells us, however, that Monsanto sent a second letter, "And it said there will be dire consequences for Fox News if the story airs in Florida." We hear that the reporters were asked to rewrite the story a total of eighty-three times, making changes they found unacceptable. They were asked, for example, to remove the word "cancer" and use only "human health implications." When they stood their ground, refusing to broadcast what they believed to be falsehoods, they were fired.[120]

Akre sued the station under Florida's whistleblower statute, which protects those who try to prevent others from breaking the law.

According to the Associated Press: "In August 2000, a jury awarded Akre $425,000, saying the station retaliated against her for threatening to blow the whistle on a false or distorted news report."

Then in 2003, "The appeals court said Akre's threat to report the station's actions to the FCC didn't deserve protection under the state whistle blower's statute."[121]

The local *St. Petersburg Times* explained: "The statute states that an employer must violate a 'law, rule, or regulation.' But the FCC's policy against falsification of the news doesn't fall under any of those categories, the court held."[122]

A policy is not a law.

DOCTOR FUN

The unfortunate side effects of excessive animal steroids

Fox then released the story "Fox 13 Vindicated," reporting that the appeals court had ruled that Akre had "[n]o claim against the station based on news distortion." The YouTube clip, however, with perhaps more attention to detail, presented the ruling on her claim as follows, "But her appeal court judges found that falsifying news isn't actually against the law."[123] Good to know.

Thanks to the strength of the dairy lobby, we have been taught, for years, that cow's milk is a vital health food. Many of us were brought up on it. And as we grew up watching Mr. Spock, the brilliant "strict vegetarian" Vulcan on TV, many of our mothers were reading the baby and childcare bibles written by his namesake Dr. Benjamin Spock. Those mothers would be fascinated to learn that in the last edition of his book before his death, Spock wrote, "There was a time when cow's milk was considered very desirable. But research, along with clinical experience, has forced doctors to rethink this recommendation."[124] He wrote that cow's milk is "high in the saturated fats that cause artery blockages," and "Dairy foods can impair a child's ability to absorb iron." He suggested that dairy foods may cause ear and/or respiratory problems, and may be linked to childhood-onset diabetes.[125] He stressed that infants should drink only human breast milk and older children should try soy and rice milk products.[126]

Spock's whole section on nutrition is terrific reading, and could be summed up with his line, "A vegetable based diet for children is generally more healthful than a diet containing the cholesterol, animal fat and excessive protein found in meat and dairy products."[127]

Chew the Right Thing

Yes, we have all heard a few media-hyped horror stories of vegan parents starving their children to death. A well-known example was the sad case of Crown Shakur, covered in Nina Planck's *New York Times* opinion piece "Death by Veganism." When seriously disturbed omnivores starve their kids, the headlines never read "Death by Omnivorism." Thank heavens for *New York Times* public editor Clark Hoyt, who took Planck and the *New York Times* op-ed editors to task for Planck's B-grade drama script about the dangers of feeding children a vegan diet.[128] Hoyt quoted a clinical nutritionist at the Children's Hospital of Philadelphia who told him that Planck's article "was extremely inflammatory and full of misinformation." He reminded us that in the pathetic case of Crown Shakur, "The prosecutor argued—and the jury believed—that Crown's parents intentionally starved him to death." Hoyt pointed out that "the baby was not given enough food to survive, regardless of what the food was" and that the Fulton County prosecutor who handled the case said it was "absolutely not" about veganism. Hoyt also quoted from the American Dietetic Association report which I cite below—a considerably more reputable source than Nina Planck.[129] As for Planck, I wonder if she is now busy at work on a *Lifetime* movie *Vegan Killers Tried to Starve My Baby.*

If you do read or see any horror stories in the media about vegan children being starved to death, please, as your heart aches for all children killed in bizarre ways by their parents, write a letter to the editor or to the show reminding people that these sad cases are not specific to vegans. (You'll read more about the power of media feedback in chapter 8 of this book.) And if you are planning to bring up a vegan baby, pick up one of the terrific new books packed with health information specifically for vegan mothers.

I noted above that unlike the Atkins diet poster boy, the founder of the Vegan Society recently died at age ninety-five. He had been in great health well into his nineties. That is not surprising given the following report from the conservative American Dietetic Association and Dietitians of Canada:

In the wake of all these health scares we've changed our message somewhat.

www.cartoonstock.com

> Well-planned vegan and other types of vegetarian diets are ap-
> propriate for all stages of the life cycle, including during preg-
> nancy, lactation, infancy, childhood, and adolescence. Vegetarian
> diets offer a number of nutritional benefits, including lower levels
> of saturated fat, cholesterol, and animal protein as well as higher
> levels of carbohydrates, fiber, magnesium, potassium, folate, and
> antioxidants such as vitamins C and E and phytochemicals. Veg-
> etarians have been reported to have lower body mass indices
> than nonvegetarians, as well as lower rates of death from ischemic
> heart disease; vegetarians also show lower blood cholesterol lev-
> els; lower blood pressure; and lower rates of hypertension, type 2
> diabetes, and prostate and colon cancer.[130]

For those vegetarian gals eating lots of tofu: In 2006, the *Journal of the National Cancer Institute* published a report analyzing eighteen epidemiologic studies published from 1978 through 2004 that examined soy exposure and breast cancer risk. The report found that "high soy intake was modestly associated with reduced breast cancer risk."[131] And a study of 97,275 women, published in the *American Journal of Epidemiology*, has found that soy foods may also lower the risk for ovarian cancer.[132] Conversely, a report on endometrial cancer from the *International Journal of Cancer* tells us that women who received the most calories from animal protein had twice the risk of the disease compared to those who took in the fewest calories from animal sources.[133]

An eight-year study of 35,000 women aged between thirty-five and sixty-nine, published in the *British Journal of Cancer*, has reported: "In our analysis, significant increased risks of incident premenopausal breast cancer in relation to increased consumption of total meat and non-processed meat were observed."[134]

When the chairman of the Scottish region of the Guild of Q Butchers heard that as little as 57 grams of red meat per day showed an effect, he pointed out that 57 grams was roughly half a quarter-pound burger and dismissed the report as "rubbish."[135] Hmmm . . . who to believe, the *British Journal of Cancer* or the Butcher's Guild?

The most comprehensive study of associations between mortality and diet found that mortality from ischemic heart disease was 24 percent lower in vegetarians than in nonvegetarians.[136]

No wonder a study published in the *British Medical Journal* showed that a higher IQ at age ten years was associated with an increased likelihood of being vegetarian at thirty![137]

Taking Care with Vegan Fare

A vegan diet is incomparably more healthful than a standard Western diet, and also a wonderful ethical choice. Should you decide to go vegan you will need to do a little research to make sure you get all the necessary nutrients and reap the benefits.

The Web site for the Physicians Committee for Responsible Medicine, www.pcrm. org, is jam-packed with information on good vegan health.

Vegan Outreach's site Veganhealth.org is also wonderfully useful. It tells us, for example, how to get our omega-3 fatty acids—they are ALA, EPA, and DHA. To get all three, we can take flaxseed as a source of ALA. Most people's bodies can convert ALA to EPA, and EPA to DHA. Because not everyone's body is perfectly efficient at the conversion, it is also a good idea to take a vegetarian source of DHA. (Like ALA, DHA can also be converted to EPA as needed, so if we have plenty of ALA and DHA we should be set.) DHA is commonly found in fish, who get their DHA from algae—and so can we, directly, cutting out the middleman, who is now too often full of mercury, PCBs, and hormones. It's great to see that Silk soy milk has started adding algae-derived DHA to its enriched variety! Veganhealth.org has plenty of information on omega-3 supplementation, as well as calcium intake guides and other vegan-helpful tips.

Protein is less of a concern than the hype would suggest. Most of us living in Western nations get more than enough. And one look at a four-hundred-pound silverback gorilla, or a well-muscled bull, should quash the idea that it is necessary to eat animal protein in order to build big muscles.

A study published in 2003 caused an uproar when the press reported that Kenyan schoolchildren who were given a tablespoon of meat every day showed better muscle growth than those given what the press erroneously called "a vegetarian equivalent."[138] Unfortunately, one of the researchers extrapolated the results to vegan children in Western nations. That was ridiculous. Unlike most Western children, the Kenyan children were protein deficient, and those in the nonmeat group were given not a vegetarian equivalent of meat, but instead a caloric equivalent that was vegetable oil. That's pure energy—not the slightest bit of protein with which to build muscle. Perhaps the results could have been extrapolated to the developed world if the children had been given the kind of real vegan equivalent most of us can now pick up at our local supermarket, or if they had been given any truly nutritious muscle-building vegetarian foods, such as nuts or soybeans.

Interestingly, the press ignored the fact that a group given the same calories in cow's milk performed no better on any of the growth measures than the group given vegetable oil, and worse on a weight-for-height score and on a mid-upper-arm circumference measurement. So much for "Milk: It does a body good."

The following, however, is important to note: "At the end of the year of supplementation, plasma vitamin B_{12} concentrations were significantly increased in children fed the Meat or Milk meal."[139] B_{12} matters! And it is only available in animal products. That is important for those of us following a vegan diet to know—we must take B_{12} supplements. Veganhealth.org, which I mentioned above, has good information on B_{12} supplementation. The need for B_{12} also has implications regarding our attitudes toward charities that give animal products, or animals, to developing nations. I will discuss that further in chapter 8.

Some people say that eating meat is natural, that we have always done it, and must continue to do it. They point to the B_{12} issue or to our "canine teeth." When I compare my canine teeth to Paula Pitbull's, I laugh that they have been given the same name because of their position in my mouth. Mine clearly are not capable of anything comparable to hers. Perhaps somewhere in Transylvania there are people whose canine teeth deserve the title, but I have never met any. To quote Harvey Diamond: "You put a baby in a crib with an apple and a rabbit. If he eats the rabbit and plays with the apple I will buy you a car."[140]

Does that mean it is wrong or unnatural to eat meat? It probably is natural. Humans are omnivores. Our ability to eat a wide variety of protein sources, including other animals (generally after cooking them) has helped our species survive. And I have already written that if it is a matter of survival I could never suggest that people shouldn't eat meat. But I cannot imagine it is a matter of survival, or even of good health, for many people reading this book.

Karen Dawn

Veggie Power

Occasionally I talk with people who say they tried going vegan but found they didn't feel good. I wonder what they were eating. Though I have noted that the need for protein is generally overhyped, I will also personally admit to being a protein addict; I live on tofu and love all sorts of fake meats—like seitan, tempeh, and Gimme Lean burgers. In this day and age it is easy to feed a protein addiction while staying vegan.

My diet serves me well. I run a few miles most days, and though I have let myself go a bit, I can still do three or four unassisted chin-ups, more than any other woman I have seen working out at my gym; more than the gals chuffing down their chicken salads at the lunch bar after their workouts.

Multi–Olympic gold medalist Carl Lewis tells us that the year after he went vegan he had his "best year as an athlete ever!"[141] And Scott Jurek, perhaps the world's greatest long-distance runner, is vegan. He is the only person to win the Badwater footrace two years in a row—it's billed as the toughest footrace on earth—and he won the prestigious Western States 100-Mile Endurance Run seven times. Sharon Gannon and David Life, the extraordinary human examples of strength and grace who founded New York's famous Jivamukti Yoga School, are also vegan. And so are countless other athletes. That doesn't mean absolutely everybody is going to run marathons or win gold medals on vegan diets. But most of us, with a little care, can flourish on them.

Baywatch bodies don't happen by accident. It seems they come from hard work and good karma. Animal-loving *Baywatch* actress **Alexandra Paul** has been vegetarian since she was fourteen. She says, "Since the beginning of my career my acting contracts have stipulated that my makeup couldn't be tested on animals, and that my characters would not wear fur. When I was twenty-six I stopped using any products tested on animals and gave up wearing leather, wool, or silk." In her thirties, on her healthy veggie diet, Alexandra competed in an Ironman Triathlon and several marathons. Now in her vibrant forties, here she is after competing in the Bonaire EcoSwim 10K Race. Wow!

Courtesy of Alexandra Paul

Vegan Health

Multi–Olympic gold medalist **Carl Lewis** tells us the year after he went vegan, "I had my best year as an athlete ever!"*

Don Emmert/Getty Images

Martin Brading

Scott Jurek is the only runner to win the Badwater Race, one of the world's most challenging, two years in a row. He also won the Western States 100-mile run seven times. He credits his good vegan diet!

Luis Escobar

Guzman

Sharon Gannon and **David Life**, Jivamukti Yoga's pictures of strength and grace, are vegan for health and ahimsa.

*Bennet, Jannequin. *Very Vegetarian* (Nashville, Tenn.: Thomas Nelson, 2001).

Alicia Silverstone and her Animal Acres friend Kirby are both vegan. Kirby finds a vegan diet gives him all the nourishment he needs to build strong muscles. And this is what Alicia has to say about her vegan diet:

> "It transformed my life. My hair, my skin, my body, everything improved. Plus I have more energy. My love of animals was the thing that initially prompted my diet change. But now I realize this kind of diet is really good for my body too. . . . Within a few weeks of switching people started to tell me I was glowing" (**Alicia Silverstone**, on going vegan).

Photograph compliments of Animal Acres and Alicia Silverstone

Karen Dawn

why Hurt anybody if you Don't Have To?

I wrote, above, that to most of us, eating a lot of meat is not even a matter of good health, let alone survival. Perhaps it soon will be a matter of survival—but not in the way we might have expected. The habit of eating meat now raises the question of our survival on this planet. Factory farming seems to be the only way to produce a lot of meat for billions of people. In chapter 7, I will discuss the effect it is having on the environment.

As for the animals, some people might argue that if we did not use them for food, they would never be born. When applied to factory-farmed animals, that argument is feeble. No person could visit a battery hen or pig confinement facility and seriously argue that the animals inside are lucky to have been born.

Perhaps we could look at that farm I love in Southern California, where the hens wander happily with their flock till their natural deaths. We could note that if humans didn't consume their eggs, the happy hens would not have been bred. But we must remember that their brothers were also bred, and are nowhere to be seen.

Moreover, current standards don't allow us to think that other "free-range" animals live like those hens. And as free-range popularity rises, those standards seem poised to fall even further. Even animals from those few farms where their lives look worth living usually face slaughterhouses that are under-regulated. Let's not forget that the overwhelming majority of animals used for food—birds, and also fish, for that matter—face slaughter with no federal regulations at all. For now, at least until "in vitro meat" becomes widely available, it is hard to deny that choosing plant-based foods is the best way to avoid contributing to suffering. Why hurt anybody if you don't have to?

For those of us living in big cities with Whole Foods around the corner, plant-based diets are easy and fun. And Internet shopping with all those delicious veggie jerkies to buy online has even made veganism fun for animal lovers and health nuts in rural areas. Staples like rice and beans are easy to come by everywhere.

Perhaps you find dining out vegan too hard. You could follow the lead of some of the Jewish kids I grew up with who respected their religion but also wanted to be able to socialize: they kept kosher at home. Why not be vegan at home? Not being able to find veggie burgers in every restaurant doesn't mean you have to have a fridge full of eggs and chicken!

If you aren't ready to "Go Vegan" even at home, there is no need to throw your hands in the air and then have a feedlot steak with battery-cage eggs for breakfast and a tortured turkey sandwich for lunch. If you are eating animal products three times a week, rather than three times a day, while that's bad luck for those animals you are eating, it is awfully good luck for the many others you are saving by not helping create demand—and it's a lot better for you than eating animals every day. I did not go vegan cold turkey, so to speak. I ate fewer and fewer animal products over the course of a few years and found that I felt better and better physically and ethically. We all need to remember that every time we choose the veggie burger instead of the turkey burger, the tofu stir-fry instead of the pork or chicken, or the soy latte instead of skim—every time we pick up our forks, knives, or cups, we make a difference.

chapter six

animals anonymous: animal testing

torture is legal

The vivisection industry likes to portray its detractors as misanthropes who prefer mice to men. Press releases depict an industry that conducts painless tests on mice to discover cures for cancer. Some people would oppose such experiments on moral grounds—it's not as if we will ever see mice strapping themselves to little tables in the hope of advancing human health. Nevertheless, if the full picture of animal testing was anything like the promotional images the industry releases, we wouldn't have the growing and increasingly vocal and outraged antivivisection movement we have today. Most animal tests are not aimed at curing cancer, or other life-threatening diseases, and those that are generally don't prove fruitful.

Neither is all testing carried out on mice. And animal testing is far from painless. In fact, there are no federal legal requirements for any care to be taken to minimize pain caused to mice and rats—the majority of the animals used in experiments. They are exempt from federal protection. While animals of all species are tortured for trivial experiments, mice and rats are tortured by the millions.

I choose the word "torture" carefully, as I think animal advocates should avoid overdramatic language. Too often we are accused of letting emotion cloud reason and we should try not to invigorate that caricature. Yet often we sidestep, too vehemently, that oh-so-dangerous hole of hyperbole and find ourselves in the murky field of euphemism. As I started this chapter on vivisection, I often noticed the word "torture" tickling my fingertips. My first instinct was to edit it away. Reluctant, however, to let euphemism cloud images of suffering, I looked up the word "torture" on Dictionary .com and noted the first entry: "The act of inflicting excruciating pain, as punishment or revenge, as a means of getting a confession or information, or for sheer cruelty." I was thereby assured that the use of the word "torture"—that is, "The act of inflicting excruciating pain . . . as a means of getting . . . information"—is an appropriate rather than hyperbolic description of animal testing, and one I will employ as needed in this chapter.

Do Animal Tests Save Human Lives?

Proponents of animal testing like to boast that no lifesaving drug reached the market without animal tests. The claim is true. At the same time, it is misleading. It implies that animal tests are part of the development process, whereas they often have nothing to do with it. Unfortunately, no drug is allowed onto the market until it has been put through a battery of animal tests.

We might have many more lifesaving drugs if it weren't for the insistence on animal testing. Thank heavens penicillin wasn't tested on guinea pigs before it came onto the market and showed miraculous results in people. At the high doses employed in animal tests—sometimes staggeringly high doses that are necessary because of the dissimilarity in the ways different species process chemicals—penicillin kills guinea pigs.

While millions of animals die in tests on cancer drugs, millions of people may die because of the reliance on animal testing. Mice are the animals of choice; they are relatively cheap, have short life spans, and are easy to control. But an article in *Nature* tells us, "Most mouse tumors originate in different types of tissues from ours." It explains that putting human tissue into mice doesn't solve the problem because "human cells are likely to behave differently in a mouse than in a human body." And it notes, "Of the potential anticancer drugs that give promising results in tests on mice with cancer, only about 11 percent are ever approved for use in people." [1]

The *New York Times* reported the following on how bad the record is with those cancer drugs that have passed animal tests—mice and other:

> One in 20 prospective cancer cures used in human tests reaches the market, the worst record of any medical category. Among those that gained approval in the last 20 years, fewer than one in five have been shown to extend lives, life extensions usually measured in weeks or months, not years. True cancer cures are still exceptionally rare. Medicines have been approved for colorectal cancer. Patients who take every one of the high-tech drugs have to spend, on average, $250,000, suffer serious side effects and gain, on average, months of life, according to studies.[2]

Those numbers make it clear: Drugs that cure cancer in mice do not work, or at best work minimally, in people. Conversely, and tragically, drugs that kill mice might have cured people. Every year thousands of people with advanced cancer, deemed terminally ill, are denied promising treatments as those treatments are tried out on animals. The vast majority of the treatments are rejected, as they harm some animal in some way. But aspirin causes birth defects in mice and rats. Ibuprofen causes kidney failure in dogs—even at very low doses. With regard to prospective cancer drugs, the *Nature* article notes: "It is also possible that drugs that would have worked in people failed in preclinical mouse trials, although there is no way of knowing."[3]

What we do know is that many people termed terminal would happily play "guinea pig" if regulations would allow it. Some would probably live.

The low correlation between species for substance tests works the other way too—substances that have no ill effect on other animals can kill us. Sheep enjoy arsenic, and strychnine is safe for a monkey in doses that could wipe out a whole human family. Or look at cigarette smoking, which has killed so many of our family members. For years the cigarette companies used research on dogs to prove that cigarette smoking does not cause cancer. Only the overwhelming preponderance of human statistical evidence shut down that nonsense. Still, today, dogs, monkeys, and other animals die in smoking and nicotine tests. As they are generally killed after the tests, not killed by the actual smoking, at least you know you have nothing to worry about when your dog enjoys a postdinner smoke.

" I don't suppose you'd like a last cigarette? "

www.cartoonstock.com

In *Sacred Cows and Golden Geese*, the doctors Ray and Jean Greek put forward a comprehensive study of the inadequacies of animal research due to the differing effects of drugs on different species.[4] One point they make is that different drugs don't only have different effects on different animals: Between 1998 and 2000, ten drugs were recalled in the United States because of side effects that occurred almost exclusively in women. ABC *World News Tonight* did a piece on the fact that aspirin, and many other drugs, affect men and women differently. It told us, "In fact, there are many common medications that have significantly different results, depending on gender."[5] It noted that blood thinners, pain medications, and sedatives are broken down differently in male and female bodies. Just like feelings and football scores.

Dr. Ray Greek comments on the male and female differences: "If a man cannot predict what a drug is going to do in a woman, what, pray tell, makes us think that a mouse is going to predict what that drug is going to do? I think a very nice existential question is, 'Which person is the mouse representing? Is the mouse representing the man or is the mouse representing the woman?' . . . Again, a man cannot predict what a drug is going to do in a woman, so this cross-species testing concept has outlived its usefulness by about a century."[6]

One of the first famous cases showing the danger of relying on animal tests occurred in the 1960s when thalidomide was released. According to *Time* magazine: "Along with the application came a sheaf of reports on years of animal testing. . . ."[7] The drug was administered as a sedative to pregnant women. It caused thousands of women to bear severely deformed children who were missing limbs.

Even after the disaster was discovered in humans, scientists could not replicate it in other animals. Finally they tried the white New Zealand rabbit and engendered birth defects at doses twenty-five to three hundred times that given to humans. Then—and only because they knew what they were looking for—they managed to produce some birth defects in monkeys given ten times the normal human dose.[8] If the researchers had not already known about the effects, the chance they would have found them in other animals is small. Remember, years of animal testing before the drug was released showed no ill effects.

Deaths from drugs deemed safe in animal tests are common. A recent example, Vioxx, appeared harmless and even seemed to protect the cardiovascular system in some animal experiments. About twenty million Americans used the drug before a clinical study on humans revealed in 2004 that it could double the risk of heart attack or stroke.[9]

We saw the subtle but critical differences between species in 2006, when six healthy young men nearly died during the trial of an experimental antibody that was meant to calm the immune system. The men's immune systems went into overdrive, attacking their internal organs. In tests, macaque monkeys had received five hundred times as much antibody as the men, with no negative reaction. *New Scientist* explains that subsequent tests have shown that the antibody "sends human white blood cells berserk when 'tethered' to a physical surface such as the base of a lab dish or in a layer of cells, as you would find in the organs of the body." However, "The antibody had previously been tested on human cells only in solution, where it failed to produce this effect. Crucially, when macaque cells replaced the human cells the tethering had no effect, suggesting that the response is unique to humans—and explaining why the monkeys in the original trials didn't get sick."[10] In other words, if the researchers hadn't bothered with the monkey tests, even at up to five hundred times the human dosage, and had instead done more in vitro tests with human cells—such as a test in a lab dish—the six men would be healthy.

First clue that the latest medical breakthrough isn't quite there yet.

Nevertheless, animal tests remain the cornerstone of research. An amusing comment by Dr. Jerry Avorn, a professor at Harvard Medical School, brings home the widespread knowledge of the ineffectiveness of the drug testing system. On an ABC special he told us that many doctors say in jest, "Always wait a year before prescribing a new drug. And if it's for a family member, wait five years."[11]

Some animal advocates do the animal rights movement a disservice by suggesting that no medical advances have resulted from animal tests. That stance weakens our credibility since we have certainly seen some surgical techniques and various treatments tested and perfected on animals. (Whether that is morally acceptable is a different matter that people can consider for themselves.) The proponents of animal testing, however, wildly exaggerate the part it plays in saving lives. That's to be expected—animal testing is big business. In the United Kingdom the Advertising Standards Authority recently ruled that the Association of Medical Research Charities could not proclaim, "Some of the major advances in the last century would have been impossible without animal research." The statement was not allowed to stand because "the advertisers could not show what would have happened if research had been carried out differently."[12] Who is to say what would have been possible in the field of medical research if the billion-dollar animal research lobby had not kept us mired in the swamp of animal testing?

Should We Refuse Animal-Tested Medicine?

Oh, please. There is no medicine not tested on animals. If we had the choice, we would take medicine not tested on animals, and we want that choice. In the meantime, how would we be helping the animals if we refused to take medicines that have been put through a battery of animal tests by a government beholden to the animal testing industry? If I were an animal who had died in one of those tests, the one person I would like to see kept strong by those medicines would be the person fighting to end superfluous animal testing.

Spare Parts—Xenotransplantation

We noted that substances that have no effect on other animals can kill people. That is also true with some viruses, which is what makes one of science's most recent endeavors so threatening, not just to the animals but also to the human race. Researchers have been working on breeding animals for spare parts when humans need organ transplants. PBS's *Frontline* series covered that issue comprehensively in a show called "Organ Farm."[13] The program explained that animals harbor viruses that do not make the animal sick, but could kill a person into whom they are transplanted. The greatest threat to human health would be if the virus were to mutate, as viruses do, into a disease that could be transmitted person to person. For example, researchers believe that AIDS most likely originated in other primates, where it did no harm, then was picked up by humans either through the consumption of "bush meat" (often monkeys or chimps) or in the development of vaccines. By transplanting organs of other animals into humans, particularly humans whose immune systems have been suppressed to allow for the transplants, we are inviting a similar disaster. If a transplanted virus were to mutate into one that is airborne, it is hard to imagine the consequences.

ANIMAL TO HUMAN TRANSPLANT PROJECT

THIS IS A CONSENT FORM

Nik Scott

Though humans are exposed to pigs every day, there are usually barriers that interfere with virus transmutation. As Jonathan S. Allan at the Department of Virology and Immunology, Southwest Foundation for Biomedical Research, explains it, "In the transplant setting, you provide the most ideal environment by introducing a virus with the organ, overcoming all natural barriers, immunosuppressing the heck out of them. So you're creating the most ideal situation for a virus to arise. . . . Transplanting whole organs obviously creates the greatest risk immediately, because it's an ecosystem waiting to happen."[14]

Hugh Auchincloss, from the FDA subcommittee on xenotransplantation, says, "We talk about recombination events, where the pig virus joins up with genetic material that's already in the human cells and makes something completely brand-new. These events can occur, we believe, and we don't know what the outcome of them would be."[15]

Bitter Medicine

Every year drug companies introduce scores of new drugs that contribute almost nothing—they replace those for which patents have expired, since drug companies make only a fraction of the money once a drug has generic replicas on the market. Hundreds of thousands of animals die each year in tests for those copycat or "me-too" drugs. For a drug to be approved by the FDA, it need not be better than—in fact, it need not even be as good as—drugs already available to the public. It just has to be proven to be better than a placebo—better than nothing. Perhaps we should be thankful that as of yet, the FDA is not letting its drug company friends patent placebos.

Dr. Sharon Levine explained for an ABC News special on the issue:

If I'm a manufacturer, and I can change one molecule and get another 20 years of patent life and convince physicians to prescribe,
and consumers to demand, the next form of Prilosec or weekly
Prozac, instead of daily Prozac, just as my patent expires, then why
would I be spending money on a lot less certain endeavor, which is
looking for brand-new drugs?[16]

According to an article in the *New York Review of Books* by Harvard Medical School's
Dr. Marcia Angell:

Of the seventy-eight drugs approved by the FDA in 2002, only seventeen contained new active ingredients, and only seven of these
were classified by the FDA as improvements over older drugs. The
other seventy-one drugs approved that year were variations of old
drugs or deemed no better than drugs already on the market.[17]

Every one of those mostly redundant new drugs released in 2002, and each one
released every year since, was tested on animals.

When Peter Jennings examined the issue on his ABC special in 2002, he chose as
examples the new, now infamous, Celebrex and Vioxx, which are drugs to treat pain
and inflammation. Jennings explained that for one hundred years aspirin has been an
effective nonsteroidal anti-inflammatory painkiller. New discoveries in the same category have been ibuprofen and naproxen, two commonly known brands of which are
Advil and Aleve. There are more than forty other drugs on the market that all do pretty
much the same thing. They were all tested on animals. A huge advertising campaign announced the arrival of Celebrex and Vioxx, and the public paid a fortune for them. But
Jennings informed us, "Neither Celebrex nor Vioxx has ever been proven to be more
effective at treating pain than drugs such as Advil or Aleve."[18]

The makers of Celebrex did claim the drug was easier on the stomach than other
drugs. But doctors from a nonprofit managed-care organization looked at the study and
concluded that the company had overstated the drug's benefits. The nonprofit even

found that people who were taking Celebrex continued to have clinically significant gastrointestinal bleeding.[19] Jennings reported that even the FDA "told the company there wasn't enough proof their drug was easier on the stomach."

USA Today, covering a study in the *Archives of Internal Medicine,* reported:

> Millions of Americans who were prescribed Vioxx or Celebrex in the drugs' first few years on the market could have safely taken older, cheaper painkillers such as ibuprofen, a new study says. . . . The drugs' main selling point was that, theoretically at least, they were less likely to cause bleeding and other serious stomach complications than older nonsteroidal anti-inflammatory drugs, or NSAIDs. But patients at low risk for such problems accounted for two-thirds of the growth in Vioxx and Celebrex use from 1999—when the drugs came on the market—to 2002. . . .[20]

© 2005, Mike Twohy. Dist. by The Washington Post Writers Group

M2Ecomics@aol.com

"Side effects include hoarding and disinterest in the exercise wheel."

I was somewhat amused by an article in the *National Review* slamming the ABC special.[21] It claimed that Jennings got most of his evidence from an "HMO- and insurance-industry-written" screed. In fact, the special cited a wide variety of sources and included interviews with Harvard Medical School doctors and the former editor of the *New England Journal of Medicine.* Ironically, the *National Review* piece sang the praises of Vioxx and Celebrex. It was published a couple of years before Vioxx was pulled from the market and the FDA requested a black box label—the most serious warning label a medication can receive—for Celebrex.

My amusement was tempered by the knowledge that people died as companies spewed their expensive profit-making treats onto the public. And countless animals died in the development of those unnecessary and even harmful drugs, which racked up some of the biggest prescription sales numbers in history.

Testing Painkillers Painlessly?

The issue of painkiller testing is prickly. People like to think of animal testing as something done while inflicting as little pain as possible. One cannot, however, find out whether painkillers work without inflicting pain. We already have painkillers that work. Few of us will ever find ourselves in life, or on our deathbeds, in unalleviated pain. Scientists might like to find us drugs that work even better—quicker, longer. But is it ethical to torture other animals to find a better version of something that already works?

What We Already Know Hurts Them

The Bayh-Dole Act provides that universities and small businesses can patent discoveries emanating from research sponsored by the National Institutes of Health, the major distributor of tax dollars for medical research.[22] Before it passed, taxpayer-financed discoveries were in the public domain. Seems fair, doesn't it? You paid for it, you get to read it. Results were available to any company that wanted to use them. Now companies can release whatever studies they like, and label others "Proprietary information." That has become a public safety issue—a company need not release the results of a study, even a taxpayer-funded study, if the results do not support its product. Dr. Drummond Rennie, an editor at the *Journal of the American Medical Association*, says that drug companies suppress results "all the time."[23] The suppression of studies is also an animal issue, as animals are killed in tests that have already been done at other companies but not made public.

The Environmental Protection Agency is undertaking a massive testing program of substances in common use, some of which are already known to harm people. Many of the substances have previously undergone extensive animal tests. Because there is no central bank of animal test information, however, the EPA is planning to start from scratch on many tests, unnecessarily killing thousands of animals.

Dr. Angell writes:

> Unlike other businesses, drug companies are dependent on the public for a host of special favors—including the rights to NIH-funded research, long periods of market monopoly, and multiple tax breaks that almost guarantee a profit. Because of these special favors and the importance of its products to public health, as well as the fact that the government is a major purchaser of its products, the pharmaceutical industry should be regarded much as a public utility.[24]

Those are strong points. Moreover, if animal testing were truly seen by society as a necessary evil, to be undertaken only when the public health benefit could reasonably be seen to outweigh the harm to the animals, then the replication of tests by different companies would be out of the question.

Necessary Evils?

We saw, above, that animals die in tests because regulations require all drugs to be tested on animals, and countless unnecessary new drugs are released onto the market every year to increase drug company profits. Animals are also used in hideous research that brings government funding to research departments at universities. How many people, if they knew about them, would think the following studies are worthy uses of their tax dollars and of the lives of animals?

Maternal Deprivation Madness

Those of us who studied psychology in college learned about Harry Harlow's maternal deprivation studies, which demonstrated the importance of maternal attachment. His studies are described in a heart-wrenching chapter in *The Monkey Wars*, a superb and evenhanded book, which is based on a series of articles for which the author Deborah Blum won a Pulitzer Prize.[25] Blum tells us that Harlow took baby monkeys from their mothers, locked them up alone in "devices that would have trapped a Houdini," and kept them there for months. She writes that it sometimes took two lab workers to hold the struggling mother monkey down while a third pulled her baby away. The babies were permanently psychologically damaged. When later placed with other monkeys, they sat staring through the bars as if alone. In one experiment monkeys were isolated for two years and "mentally destroyed." No amount of company, or stroking, or any interaction could even get them to lift their heads.

Photograph compliments of PETA

Harlow found that the baby monkeys would cling to stuffed surrogates placed in their cages. And he found that even when those surrogates blasted babies with air, or stabbed the infants with spikes, the babies still clung. Then he found that if he later inseminated those abused monkeys, they tortured their own babies—slammed them against walls or chewed off their fingers.

Lest anyone think that Harlow was distressed by what was happening to the monkeys, here is a quote from the man who had so many of those beings in his charge:

> The only thing I care about is whether or not the monkeys will turn out a property I can publish. I don't have any love for them. Never have. I don't really like animals. I despise cats. I hate dogs. How could you like monkeys?[26]

Reading descriptions of Harlow's work is harrowing. It may be fair to say we learned something about parenting from it—though surely the monkeys would say at too high a price. But once we got the gist, you'd think a civilized society would insist they stop.

No. Fifty years later scientists are still carrying out maternal deprivation studies. Blum shared a sad joke:

Q: How many monkeys does it take to convince scientists that baby animals are stressed by being taken away from their mothers?

A: As many monkeys as the government will buy for them.[27]

These days the government particularly likes to buy monkeys for tests on life-threatening illnesses. Researchers find it easiest to get funding for AIDS research. So we see studies that have "simian acquired immune deficiency syndrome" in the heading, which discuss the effects that occurred when "animals were relocated from their natal cages to indoor housing in standard individual laboratory cages."[28] They are maternal deprivation experiments renamed. Those who give to AIDS research charities might wish to take note.

The old-fashioned maternal deprivation studies that don't even pretend to deal with AIDS are still going strong. They have consumed thousands of monkeys and tens of millions of taxpayer dollars. From 2000 through 2003, Oregon Health & Science University's Judy Cameron received $8,792,889 in National Institutes of Health grants.[29] A 2005 Associated Press article covered a study in which "Cameron took rhesus monkey mothers from their babies at the age of 1 week and 1 month." It reported, "'What the study shows is that trauma early in life has long-term consequences,' said Cameron."[30]

Is there anybody reading this book who didn't know that?

Baboons on Ecstasy

People might be surprised to learn that the taxpayer-supported National Institutes of Health also funds primate ecstasy experiments. The Drug Enforcement Administration lists ecstasy in the "schedule 1" category, meaning it has no medical value and carries serious risks. In 1998, researchers at Johns Hopkins University, in tests on human habitual users, demonstrated that it causes long-lasting nerve cell brain damage.[31] Yet in 2002, the same researchers did more tests, this time on monkeys and baboons. They reported in the journal *Science* that the animals were given two shots of ecstasy. Two out of ten died after two shots, and two more were in such distress that they were not given a third shot. All of the animals were killed a few weeks later so that their brains could be examined.[32]

Despite numerous attempts, the researchers were unable to replicate the study and get more baboons to die in immediate distress from being given ecstasy. Finally, a year later, the researchers announced that in the original study, oops, the drug vials had been mislabeled. They had given the primates large doses of methamphetamine, commonly known as speed, instead of the similarly named methylenedioxymethamphetamine, or ecstasy.[33] Understandable, really—I'm sure we all make that mistake occasionally: the recipe calls for ecstasy but, darn, you pick up the shaker of speed.

Seriously, while the effect of speed and ecstasy on the brains of other primates may be a topic of interest, can a moral society allow scientists to torture and kill in order to get more detailed information about recreational drugs that have already been proven dangerous and been banned? Surveys suggest that the vast majority of the American public, though not comfortable with animal testing, accept it as a necessary evil. Surely ecstasy experiments cannot count as necessary. To many of us, the thought of going to work every day and killing baboons with speed or ecstasy seems so unnecessarily evil it is hard to imagine anybody not evil could do it, and that a civilized society would allow it. Ours does.

He's Just Not That into Ewe

The tennis great Martina Navratilova has come out against the Oregon State University and Oregon Health & Science University's research designed to change the preferences of homosexual animals. According to the researchers, about 8 to 10 percent of male sheep prefer same-sex partners. The *Weekend Australian* reported, "Researchers are trying to adjust the rate at which estrogen is absorbed into the brain to see if that will make the 'gay' rams become attracted to ewes."[34] The rams are killed at the end of the study so that their brains can be examined.

Navratilova's celebrity status has helped bring attention to the matter—a welcome change, as the shenanigans of university researchers rarely get public scrutiny. It has been fascinating to watch the performances of the gay-ram researchers since their studies were outed by Navratilova, backed by PETA. They publicly accused PETA of lying about the studies, particularly about any implications for humans.

I have read the Research Protocol Application, however, which states, "The primary goal of the proposed research is to understand the neuroanatomical and neurochemical basis of sexual preferences in male sheep." It notes that the research "has broader implications for understanding the development and control of sexual motivation and mate selection across mammalian species, including humans."[35]

A *New York Times* article on the controversy includes quotes from one of the researchers attempting to back away from that statement. The *Times* article says:

> Mentioning human implications, he said, is "in the nature of the way we write our grants" and talk to reporters. Scientists who do basic research find themselves in a bind, he said, adding, "We have been forced to draw connections in a way that we can justify our research."[36]

So let's get that straight, so to speak: We shouldn't believe what the researchers wrote about human implications, because we should understand that they would write anything in order to get a research grant, but we should believe them now as they face public outcry about what they wrote. Uh-huh.

If there are no human implications, why are sheep being killed in experiments designed to change their sexuality? Surely this is not the vital lifesaving research for which the biomedical community tells us we must, oh so regrettably, sacrifice animals' lives, while giving thanks for the great contribution they have made to human survival.

The upside of the fiasco has been some of the publicity. The *Weekend Australian* article, for example, includes enlightening information for those inclined to view homosexuality as unnatural. It discusses homosexual acts that have been documented in kangaroos, emus, fish, foxes, seagulls, and whales. It even lets us know that among swans, so often held up as advertisements for lifelong monogamy, about one in five pairs are of the same sex.

Navratilova has written to the university and implored, "For the sake of the animals who will die unnecessarily in these experiments and for the many gays and lesbians who stand to be deeply offended by the social implications of these tests, I ask that you please end these studies at once."[37]

PETA comments, "The trials can only be described as a needless slaughter of animals, an affront to human dignity and a colossal waste of precious research funds."[38]

Ironically, PETA is often accused of being on the fringe, but their thinking on that one has to be mainstream. Surely a ban on same-sex animal marriage should be enough.

Femme Fatalities

Really Painful Periods

Undercover video taken in 2003 in the Columbia University primate lab (available at www.FemFatalities.com) shows a baboon bleeding from a wound in her head where a metal pipe had been inserted in order to observe the effects of stress on her menstrual cycle. Seriously. How many women who have had trouble with painful periods would be glad to know the researchers are doing that to try to help?

. . . And Difficult Pregnancies

Researchers have been conducting experiments to find out if high doses of vitamin C might counteract the negative effects of smoking during pregnancy.[39] If so, pregnant mothers might be able to continue to kill themselves, while safely preparing to bring new lives into the world. Indeed, the researchers argue that the studies are important because despite the wealth of evidence about the effects of smoking on unborn children, they cannot persuade all of their pregnant patients to quit. So they make monkeys match the women's hurtful behavior, and then kill and examine the monkey babies. Press articles, reporting the findings, included quotes from the researchers but none from anybody worried about the ethics of the study. No press reports questioned whether such experiments fall into the category of the unfortunate but absolutely necessary experiments on other species that many people think we must do for the sake of human health. And nobody asked whether women who cannot be persuaded to stop smoking while pregnant are ready to care for babies.

In this country, every human primate has a right to have a baby and to poison that baby with nicotine if she so chooses. No nonhuman primate has the right to defend her baby from humans who wish to poison her with nicotine.

The feminist icon Gloria Steinem has sent a letter to the director of the NIH'S Office of Research on Women's Health protesting the millions of federal dollars spent every year on experiments that are cruel to animals while "taxpayers are cheated of an effective use of their dollars." She cites the nicotine experiments on pregnant animals and notes, "At the same time, there is no money for many of the pregnant women desperately in need of the help of federal under-funded programs in ending their addictions."[40]

"I don't know anybody yet who has died from PMS, so for me that is a particularly weak case to be making for the torturing of little creatures" **(Bill Maher).**

From *Behind the Mask* (2006), documentary produced and directed by Shannon Keith.

Courtesy of HBO

Nellie McKay says, "The reason that I work so much with animal rights is I feel it has the farthest to go, even though I'm very committed to women's rights and civil rights and the fight against poverty. Animal rights is still considered a joke to many people, and it isn't a joke at all."*

And "It's all the same fight. If you waited to pursue women's rights until you had complete and utter civil rights, then women would still be barefoot and pregnant. If you waited to pursue civil rights until you had workers' rights for white people, there would still be complete segregation or slavery. I don't feel you can wait or prioritize, and I don't feel that they're exclusive at all."†

* Quoted by Richard Harrington, "Nellie McKay's Ongoing 'Head'-Ache," *Washington Post*, January 20, 2006, p. T06
† "Promotion Sickness," *Mother Jones*, February/March, 2006.

Foods That Fatten Drug Company Wallets

When Michael Moore released his documentary *Sicko*, exposing the weaknesses of the United States health care system, PETA sent him a letter in which they congratulated him for his good work on the film but noted, "There's an elephant in the room, and it is you." Their point was that his film missed noting that the standard American diet is largely responsible for the strain on our health care system.

According to the *New York Times*, 60 percent of adult Americans are overweight and a third are obese.[41] A report done by the federal Centers for Disease Control and Prevention says obesity is rapidly overtaking smoking as the leading avoidable cause of death.[42] Who knew the porky kids from gym class would finally find a race they can win? They're rolling right pass the class-cutting smokers on the race to the grave. Obesity is actually a serious issue. But how many of us have serious doubts about what usually causes it? Yet, just think, when you buy your next caramel mochaccino with whipped cream, a percentage of your taxes will kill animals to find out why you are getting fat..

The torture and killing of millions of animals in attempts to find drugs that cure obesity-related ailments really is one of the saddest ironies. We know that in the vast majority of cases, obesity results from eating too much of the wrong foods—and, as it turns out, animal products are the wrong foods. Your own personal survey might lead you to notice that you don't know many fat vegans. A huge study of fifty-five thousand healthy women found that vegetarians and vegans had a significantly lower risk of being overweight or obese than did omnivores.[43]

Hundreds of studies, of humans, have shown the lower incidence among vegetarians of such obesity-related diseases as diabetes, high blood pressure, and heart disease. (See "Ronald McDonald Strikes Again" in chapter 5 for more on that.)

Despite the wealth of human data on plant-based diets and health, the federal government grants millions of taxpayer dollars every year to animal research on obesity. After all, it's a lot better for their drug company supporters if we don't change our diets and they can sell us pills. But it probably isn't your dog who is going to be gobbling down diet pills because summer is coming, and sometimes those pills, judged safe in misleading animal tests, end up killing some of us—think fen-phen in the mid-1990s.

A *New York Times* article reported that an estimated two hundred possible obesity drugs are now in the research pipeline.[44] Meanwhile the government continues to subsidize the meat industry. The nutritionist Marion Nestle said of her days serving as editor for the *Surgeon General's Report on Nutrition and Health*, "The extent of the industry lobbying was extraordinary. On my first day on the job, I was told that no matter what the research showed, we should never say 'eat less meat' or 'eat less sugar' but use euphemisms like 'choose lean meat.' I was pretty shocked."[45]

The U.S. government subsidizes foods that kill us, and drugs that help alleviate their effects. It's as if, having found out that smoking causes cancer, instead of urging people to quit they sought treatments to alleviate the effects of smoking while increasingly subsidizing the cigarette companies. Come to think of it, that almost describes the nicotine pregnancy tests cited above, doesn't it? The whole thing would be funny if millions of people and animals weren't dying as the legislators do favors for the industry lobbyists who elect them.

Doug Hall/PCRM

killing animals to kill people

Weapons of Dog Destruction

Animals have been shot, bombed, poisoned, gassed, and killed with grotesque viruses in military experiments. In 2002, we saw an interesting scenario when CNN aired al-Qaeda videotapes, which the network obtained in Afghanistan, of dogs being killed by gas.[46] The government, journalists, and public were outraged, and the tapes aggravated anti-al-Qaeda sentiment. One wonders why CNN doesn't air tapes of our own government's experiments on animals.

A 2006 story from the United Kingdom reported that the number of animals used in military experiments, including biological and chemical warfare tests, had doubled in the last five years. A spokesperson from the British Union for the Abolition of Vivisection commented,

> If these experiments were conducted with one eye on Iraq, it's bitterly ironic that the only victims of weapons of mass destruction in this conflict turn out to be animals.[47]

Taser Tales

ABC's Denver affiliate showed video, obtained under the Freedom of Information Act, of a Department of Defense study that used fully conscious pigs to test the effectiveness of various Taser devices. Tasers are used by governments to control people without killing them—for example, on unarmed criminals or against protestors. Perhaps the Department of Defense is worried about Orwellian revolutions on animal farms.

Without doing long-term physical damage, Tasers administer agony. On the tapes we see pigs topple and convulse—or run and try to leap over the barriers of their pens. Though there is no audio with the tapes, you see the animals' mouths wide open—they are obviously shrieking.[48] Tapes of all eleven tests are on YouTube.[49]

The ABC reporter announced that the study was "in compliance with the Animal Welfare Act" and that "the Institutional Animal Care and Use Committee also gave its approval." With that in mind, you might look at the video on YouTube in order to learn the safeguards against pain offered by those bodies—none. The reporter also noted, "The study could not have been safely conducted on humans, according to officials at the Brooks City-Base." Who decides that it is okay to conduct such studies on other animals? Is it assumed that the majority of us want our tax dollars spent that way?

Unfortunately for the pigs, none died during the tests we saw on the tape. We are told that they lived—to be used in further studies. Then they were all "euthanized" by the government. Why? For that kind of service I would expect retirement in the lap of luxury.

Pigs of War

In the *New York Times*, a young medic explained that for his pre-Iraq training he had to work with live tissue. He had to learn what happens when an already wounded creature receives more wounds. He was given a pig to keep alive. He said, "Every time I did something to help him, they would wound him again." He got more specific: "My pig? They shot him twice in the face with a 9-millimeter pistol, and then six times with an AK-47 and then twice with a 12-gauge shotgun. And then he was set on fire. I kept him alive for 15 hours. That was my pig."[50]

Innocent Until Proven Animal

Animals are used in tests for capital punishment methods, such as the electric chair. When capital punishment via electric chair was reintroduced in Tennessee in 1999, a report broke that dogs had been tested in preparation for the condemned humans. That particular report was officially labeled a hoax—the specific details regarding people who were supposed to be involved were erroneous, and it is questionable as to whether any dogs were really used to test equipment during that period.[51] But the gist was right. Animals are used in tests on capital punishment devices, and the animals aren't just killed outright—testing means experimenting with doses. A *USA Today* article on the development of the electric chair reported, "Electrical pioneers such as Edison got into the act, carrying out experiments on animals to test the effectiveness of different dosages of current."[52] The *Cincinnati Inquirer* reported: "One infamous demonstration took place in a packed college lecture hall. A dog was electrocuted with Edison's DC—but survived, in agony. It took a jolt of Westinghouse's AC to kill the poor creature."[53]

Any method of killing people will first be tested on animals, who have no rights, and who, to the government, don't matter. Many people reading this book may oppose capital punishment because mistakes are made, because it is meted out under a system with heavy class and racial biases, or simply for the belief that violence is not the way to prevent violence. Others, however, might support state-sanctioned killing of people for heinous crimes, either as a deterrent or because they believe death is an appropriate punishment. How do those people feel about the torture of innocent animals in order to make the deaths of the murderers—those whom society intends to punish—as painless as possible?

Looks That Kill

In the European Union, animal testing for the development of cosmetics has been banned. No cosmetic may be sold if the finished product or any ingredient was tested on animals, provided validated alternative testing methods exist. Complete bans are expected to be in place by 2009.[54] In the United States, however, the cosmetic animal-testing industry is alive and in fabulous shape. Whereas animal testing is legally required for medication, and for any household products that have insecticidal or antibacterial properties, it is not required on cosmetics. But in the land of the lawsuit most companies decide it is better to be safe than sued. Even

Bizarro © Dan Piraro/King Features Syndicate

though companies might know how little predictive value some of the tests have, when faced with a potential lawsuit, they like to be able to demonstrate that they covered their bases.

Blinding Beauty

The Draize Eye Irritancy Test is still standard. Rabbits are used because they have no tear ducts, which would wash out or dilute the chemicals—and also ease the pain. On an episode of David E. Kelley's drama series *Boston Legal*, the character Bethany Horowitz, a lawyer representing an animal rights activist, described the process accurately while cross-examining the owner of a cosmetics company:

> You lock them in stocks so that just their heads stick out. You clip their eyelids open and pour chemicals into their eyes and they are left there for two weeks to experience ulceration, bleeding, and massive iris deterioration.
>
> Do you not subject these animals to excruciating pain? . . . Sometimes the rabbits break their own necks trying to escape.[55]

© 2005 by Linda Frost

In an article published in *Scientific American*, Thomas Hartung and Alan M. Goldberg (Goldberg is a professor of toxicology at Johns Hopkins University), discuss alternatives to the Draize Test. One example is the use of fresh eyeballs from slaughterhouses. In another alternative, the fine membrane separating the yolk of a hen's egg from the albumen is successfully substituted for the cornea. Hartung and Goldberg also explain that thanks to research with human corneal cells, "several companies have now produced three-dimensional tissues that very accurately mimic the outer surfaces of the human eye—allowing experimenters to check not only for irritation but also for subtle structural changes."[56]

Draize alternatives suffered a setback in the 1990s when the alternative tests gave results that did not match the Draize. It has since been discovered that it was the inaccuracy of the Draize Test, not of the humane alternatives, that led to the discrepancies. Efforts are again being made to validate the alternatives. The Draize Test, however, while banned in the European Union, is still widely used in the United States.

Skin Deep

To test skin care cosmetics in the United States, immobilized animals are shaved and doused with products to see how much it takes to make their skin burn and blister. Such cosmetic tests are already banned in the European Union when alternatives exist—they will be banned outright from 2009. The *New Scientist* tells us that because of the ban, companies have been forced to come up with alternatives. It reports on the release, by L'Oréal, of Episkin, a reconstructed human skin that has been approved for testing of cosmetics.

Episkin layers are grown on collagen, using skin cells left over from breast surgery. For the product to be validated, the researchers had to prove that they could produce results as effectively as with animal tests. But L'Oréal's scientific director says that independent tests have shown that in some cases Episkin is able to predict more accurately than animal tests how a person will react to products.

The article tells us: "For example, it can be adapted to resemble older skin by exposing it to high concentrations of UV light. Adding melanocytes also results in skin that can tan, and by using donor cells from women of different ethnicities, the team has created a spectrum of skin colours which they are using to measure the efficiency of sunblock for different skin tones."

Kathy Archibald of the anti-vivisection group Europeans for Medical Progress is quoted: "This is a great advance—not just for animals but for people who will finally have a safety test that is relevant to them."[57]

How long do you think it would take for Episkin, or something like it, to become the gold standard cosmetic test in the United States if animal tests for cosmetics were banned?

Botoxic

Years of hard work by animal advocates led to a ban on cosmetic testing in California. Botox, however, is produced in California, and every single batch is tested on mice, who are put through LD50 tests (see below). We may all associate Botox with happy Hollywood faces free of frown lines, but in a small percentage of cases it is used for medicinal purposes, for example, to treat migraines—caused in Hollywood by worry about frown lines. So the manufacturers skirt California's cosmetic testing ban. Money talks louder than law.

who would you kill for brighter whites?

One could only imagine a Stepford wife killing for brighter whites. But all of us do just that when we buy new improved detergents from the major mainstream companies—those that have not stopped animal testing. The barbaric LD50 test, popularized in the era of World War I, is still widely used even in the twenty-first century. LD stands for lethal dose; 50 refers to the percentage killed by the test, which involves calculating the toxicity of a substance by administering it to a group of animals to see how much is needed to kill half of them. If animals are force-fed, for example, the equivalent of a full bottle of detergent in a day (more than any person could realistically consume in one session) and none die, proving that the product is safe at least to that species of animal, they or others will nevertheless be force-fed more—whatever amount it takes to kill half of them. Death by poisoning with only moderately toxic chemicals is long and painful.

The idea sounds like it comes from one of Stephen King's horror plots. In fact, in 2003, a thriller called *Lethal Dose* (starring "Scary" Spice Girl Melanie Brown) based its premise on the gruesome tests. But LD50 tests are not horror fiction. They are happening in laboratories all across America right now. The EPA requires animal testing for all household products that make antibacterial claims.

Let's not forget that different species respond differently to various chemicals. A study that examined fifty chemicals found that LD50 animal tests predicted human lethal doses with a correlation of 65 percent—just a little better than a coin toss—whereas human cell line tests predicted human lethal blood concentration with a correlation of 74 percent.[58]

Michele Rokke, a now retired PETA undercover investigator, was employed as an associate technician at the New Jersey Huntingdon Life Sciences laboratory for eight months. In the award-winning documentary *Behind the Mask* she tells us of the work with which she was involved:

Testing methods are generally confined to a handful of well-known tests: Oral gavage, where a tube is shoved down the throat. Nasal-gastric, where the tube is shoved up into the nose. There are derma-brasion tests where substances are applied on the animal's skin after it has been shaved and abraded. Substances can be put into the eyes, up the rectum, up the vagina, injected intravenously and inhaled. Some of the worst tests to see are soft fluffy little white rats that are shoved into containers the size of a hair spray can, and they are left just hanging there, sometimes for more than a day, inhaling some toxic substance.[59]

And she tells us the story of James:

While there were so many animals I wish that I could have gotten out, who I really got close to during my eight months there, if I could have saved anybody, I wish I could have saved James. James was a primate who was so unlike the other primates there in a lot of respects. Some are bred, either nationally or internationally, and some are actually still wild-caught. James came from China, through a broker in North Carolina.

Immediately when I met him, when I went in to clean, I knew he was different from the other primates. He just really connected with me and I connected with him. I remember feeling, when I left work that day, like I finally had a friend in the laboratory—which probably sounds corny on some levels. But the other primates are obviously terrified. James was too—but the other primates would run to the back of the cage, and shriek and try to get away from people. As soon as I just gave James a few seconds of my attention outside of his cage, he reached out for me right away. It was as if he was really accustomed to people.

Over the course of the next several weeks I spent a lot of time with him—as much as I could with the busy workday. He would eat orange pieces out of my hand. And on more than one occasion when I was kneeling in front of him, he would reach out and start grooming my eyebrows and start tugging at the bonnet over my hair. He wanted to perform what his species performs, which is the natural grooming of family members.

His behavior changed so radically when the testing started. When he was first acclimated to the ECG board he became much more submissive and fearful, and as the weeks progressed it was harder and harder to get him to even focus on me. He would be sort of scattered and looking around at different things, because certainly I was a person who had compromised his trust after that point. I looked like every other technician who came and did terrible things to him. But when I see him—(Michele is watching video of him, as she talks to the camera)—it is striking to me how profound this relationship was with this little guy. And while there were so many animals that I wish I could have gotten out. . . .

At this point Michele can no longer speak because she has started to cry. We see on the screen a picture of a beautiful little primate face, and the caption "James. Killed for household product testing."

Those who make money from household products might try to convince us that James did not die in vain—after all, we probably gained some great new laundry detergent. I would suggest that James's sad life and death were not wasted only because Michele Rokke and the filmmaker Shannon Keith have immortalized him. His tale will surely galvanize the fight against product testing on animals and eventually save millions of other innocent beings from his fate.

Karen Dawn

Are Rats and Mice Animals?

In its press releases, the vivisection industry likes to discuss the strong government regulation of its work. I hope the few examples provided above have led people to question how much that regulation helps the animals. Moreover, there is something missing from all those reassuring press releases and quotes: Mice and rats are not animals. At least not according to the Animal Welfare Act, which lays down basic (pitiful) cage requirements and which established the Institutional Animal Care and Use Committees that are supposed to improve laboratory animal welfare. The USDA Web site tells us:

"The term 'animal' includes specific species in some, but not all, situations and specifically excludes rats of the genus Rattus and mice of the genus Mus as well as birds used in research."[60] Thus rats and mice are exempt from the Animal Welfare Act—even though they make up approximately 95 percent of the, um, animals used in research.

We can understand why the animal-testing industry does not want rats and mice included under the Animal Welfare Act. Their inclusion would have an enormous financial impact. The cost of housing would be just one example. Perhaps we can also understand how the industry got away with it. Rats in particular are not popular. Mickey Rat would never get parents lining up their kids for the Magic Kingdom; even the mouse had to have a fair bit of California cosmetic surgery to get the job. But rats are just as sentient as any other animals. When given the chance, rats make wonderful pets, intelligent and affectionate.

The neuroscientist Joseph LeDoux says, "There's lots to learn about emotion through rats that can help people with emotional disorders."[61]

Doug Hall/PCRM

That would not be so if they did not have emotions like ours. On that note, I must share the comments of one of the early prominent researchers in the field of emotion, Professor Calvin Springer Hall Jr. A 1939 edition of *Time* magazine reported:

> To those who protest that rat emotions are not to be compared with human emotions, Dr. Hall replies that human psychology has evolved directly from animal psychology—and that if you do make such a protest "you are not really an evolutionist, and therefore your views deserve little serious consideration."[62]

Actually, one need not be an evolutionist to acknowledge animal emotion and see the need to treat animals with mercy. But Hall does a beautiful job of challenging the historical use of science as a cloak for cruelty.

Science Daily has reported on research at the University of Georgia demonstrating that rats have metacognition, which for decades was thought to exist only in humans and was recently shown in monkeys and dolphins.[63] Metacognition is the ability to reason, or think about one's own thinking, for example to know what you do and don't know. In a test, rats were given a large reward for a correct answer, no reward for an incorrect answer, and a small reward for no answer. When the answer was not clear—for example the rats had to decide whether a tone was of long or short duration when the tone was somewhere in the middle—rats would give no answer rather than risk giving wrong answers, suggesting they know what they don't know.[64] A *Chicago Tribune* story on research into animal emotions noted, "Laboratory rats have been shown to chirp delightedly above the range of human hearing when wrestling with each other or being tickled by a keeper—the same vocalizations they make before receiv-

Activist Kamy Cunningham's pet rat, Basil.

© 2004 by Kamy Cunningham (from *The Wholistic Rat*)

ing morphine or having sex."[65] The *New York Times* tells us, "Rats dream as we dream, in epic narratives of navigation and thwarted efforts at escape."[66] Another *Times* article has reported, "Rats may be more caring and selfless than their reputation suggests. Or at least they can be very kind to each other, even to rats they have never met before." It describes a study in which rats learned that when they pulled a lever, a rat in the adjoining compartment received food. They pulled the lever significantly more often when another rat was next-door than when the next compartment was empty. The reporter, Nicholas Bakalar, comments: "Incidentally, these rats were not the usual cute, pink-eyed white lab rats. They were bred from wild *Rattus norvegicus*—the brown or gray Norway rat depressingly familiar to residents of many American cities. Is it time to stop using the word 'rat' as an insult? Maybe. Apparently even a nasty-looking rat can be possessed of sterling character."[67]

A cancer researcher has discussed the difficulty of scientific data that can become polluted because "mice are so sensitive. Their blood pressure easily goes haywire. They're like babies in that respect."[68]

Yet in laboratories rats and mice are treated like things. These intelligent and sensitive animals are "stored in racks of clear plastic containers," each one equipped only with "pine shavings, pellets and a horizontal bottle of water."[69] And absolutely anything can be done to them for any reason.

When the amendment that formally denies rats, mice, and birds any form of protection under the Animal Welfare Act was included in the 2002 Farm Bill, those lobbying for it assured legislators that adequate protection was already in place. The *Washington Post*, however, reported on an undercover video shot by People for the Ethical Treatment of Animals at the University of North Carolina:

> The video seeks to undercut that assurance. In one scene, a researcher cuts open the skulls of squirming baby rats to remove their brains without first numbing the animals in a bucket of ice— a shortcut that the researcher concedes on tape is a violation of the experimental protocol. "I don't put them to sleep," he tells the undercover technician. "Maybe it's illegal, but it's easier."[70]

Like a Used Tissue

The New England Anti-Vivisection Society Web site includes the following account from a member of an Institutional Animal Care and Use Committee:

> . . . (the researchers) made an incision in his neck and . . . inserted a catheter in his heart. The tube was passed behind his ear to keep the rat from going after it. Now the anesthesia was stopped and the rat was turned onto his stomach. A plastic dome . . . was put over him. The tube to his heart passed out a hole in the dome, and the rat was left for the anesthetic to wear off . . . but his back feet and legs were taped down, because he'd try to get away. Then it was time for the heart attack. . . . First a paralyzing drug is shot into the catheter . . . and then a drug to cause a heart attack. The injections were done quickly . . . but death was not quick, nor was he paralyzed completely. . . . He jerked many times and his head turned from side to side. . . . Tears started to run down my face. I wanted to take the little guy and gently bury him. I keep seeing that poor helpless little creature trying to escape, twitching in pain, and lying there discarded like a used tissue.[71]

Depressing Depression Studies

There are countless ways to try to measure the intelligence of an animal. No wonder a *New York Times* writer labels as "fiendish" an IQ test known as the Morris water maze, in which mice have to find a platform located within a pool.[72] Imagine having your IQ checked by a test in which you would drown if you scored low. You might be thinking that would be an excellent solution to the world's overpopulation problem—starting with your school or workplace—but we probably all agree such tests would involve an unnecessary and unacceptable level of cruelty. Mice and rats, however, are used in even more fiendish tests to measure depression. A *New York Times* article described some of the tests "that you could never do with people."[73] We read, "In one popular test, mice are placed in a pool of water and monitored to see how long they swim before giving up. If they are treated with an antidepressant, they swim longer." In another test "mice are taught to avoid an electrical shock by pushing a lever. When the lever is inactivated, the mice continue to push it anyway, even though they still get shocked. Mice under the influence of antidepressants keep pushing it longer."

That article quotes the president of research and development at Wyeth telling us that mice do not have emotions like ours. Yet the same drugs that make a mouse swim longer alleviate depression in humans. How could animal experiments lead to treatments that affect human emotion if mouse emotion is dissimilar? And if it is similar, how can we justify watching how long a mouse will swim in a pool of water before he gives up? I bet you can think of countless ways to test emotional resilience that don't involve electric shocks or drowning.

The sad irony is that researchers want it both ways. They hold that animals are enough like us for experiments on their bodies to be truly useful, but so unlike us that we needn't worry about the plight of the animals being used. Almost the exact opposite appears to be true. Because other animals differ from us in so many ways—particularly in the ways they process chemicals—animal experimentation has generally not proven rewarding in vital fields, such as cancer research. Yet animal research works in the field of emotion, in studies such as those described above for depression-alleviating drugs, because other animals are so very much like us emotionally. Knowing that, how can we justify what we do to them?

"We're also benefitting our own species in the event they start marketing cosmetics to rats."

Is It Ever Right to Experiment on Animals?

Let's Look at Aziz

Whether or not it is ever right to experiment on animals is something about which reasonable, even compassionate and ethical people, could disagree. Some animal rights activists would say no, because a nonhuman animal can never give us her consent. They would argue, reasonably, that might doesn't make right, and the fact that we can doesn't mean we should.

Other people might argue, also reasonably, that most people, even many animal advocates, would find it permissible to eat an animal if one were actually starving. (And that might have nothing to do with species—not all of us found the story *Alive,* in which survivors ate the frozen corpses of fellow crash victims, all that much more disgusting than the average Sunday barbeque.) If we can eat other beings when starving, why then would it not be permissible to kill an animal for the sake of saving oneself from lethal diseases?

As previously noted, the vast majority of animals who die in tests are not involved in tests designed to cure deadly diseases. Even those used in such tests frequently die in vain. According to an article published in the *British Medical Journal,* the studies are often poorly designed.[74] The usefulness of the animal model is being increasingly questioned.

Yet sometimes it does appear that tests done on animals have furthered knowledge that has helped, even saved, humans. An example held up as such, and widely publicized in the last few years, is the work of the brain surgeon Professor Tipu Aziz. He uses monkeys for tests in the study and treatment of Parkinson's disease and multiple sclerosis. Aziz has made news as an outspoken advocate for animal testing, and also thanks to comments by Dr. Peter Singer, the author of *Animal Liberation,* in a BBC televised interview. Singer said that assuming Aziz's tests are as helpful as Aziz has represented, they are examples of animal research Singer could see as justifiable.[75]

Facing Felix

Adam Wishart spent a year studying both sides of the vivisection debate for his BBC documentary, *Animal Testing: Monkeys, Rats and Me.*[76] As part of the film, Wishart followed a monkey named Felix as he was prepared for an experiment in Aziz's laboratory. He wrote about the experience in an article for London's *Daily Mail.*

In the article, Wishart makes it clear that he has not come to the laboratory as an opponent of animal testing. He writes that he believes that medicine has benefited enormously from experiments on animals and that drugs tested on animals extended his father's life. Yet meeting one of the animals clearly has an impact. He writes:

> It's impossible not to imprint human characteristics onto Felix's face: he has sad eyes, the wrinkles of an old man, and he frequently fixes me with an intense gaze, curious about this new visitor. I find it profoundly disturbing.[77]

Wishart explains that Felix is learning to touch a computer screen in response to various cues. Wishart watches a training session and describes the affection between the trainer and Felix: "It's like a dedicated parent trying to teach a toddler to hold a fork."

Wishart learns that electrodes will be planted in Felix's brain to monitor the electrical activity for each movement. Then a toxin will be injected into Felix's veins, inducing the symptoms of Parkinson's disease. Aziz assures Wishart that when the toxins are pumped into Felix's body "he won't really be in the same distress as a human patient." Aziz and his team will see if they can relieve the symptoms by passing electrical impulses into Felix's brain. If so, Aziz will start human trials of the same technique.

Bizarro © Dan Piraro/King Features Syndicate

Whether or not the experiment is successful, regulations demand that Felix, who will be "scientifically impure," will be killed.

We read that Aziz claims his techniques have improved the lives of forty thousand people around the world, including that of a thirteen-year-old boy named Sean, who Wishart watches on the operating table. Sean is expected to recover many of his motor skills over the next twelve months. Yet Wishart writes, "even knowing this, watching Felix in the lab and imagining what was about to happen to him made me queasy."

Toward the end of his thoughtful article, Wishart writes, "In my heart, I find the idea of Tipu Aziz and his colleagues in scrubs, hovering over the anaesthetised monkey, difficult to bear. Felix will spend months living with the electrodes in his head, performing his touch screen tasks in his tiny cage. And after that, he will be destroyed." Yet Wishart also writes that he cannot oppose the experiment on Felix or any other animals, having met Sean. He explains, "The raw emotion on his face, and that of his mother, as he struggled to stand up is a sight that will stay with me forever."

On reading the touching line about Sean's mother, I found myself wondering whether Wishart's profound discomfort over Felix's fate might have crossed the line into opposition to that fate if he had been faced with Felix's mother. As with Harlow's monkeys, did it take two attendants to hold her down, screaming, as a third pulled away her baby and carted him off to the laboratory?

I noted above that Aziz's work is held up as an example of justifiable research, an image helped by the comments of Peter Singer, who said that if it were true that Aziz's work had helped forty thousand people: "Well, I think if you put a case like that, clearly I would have to agree that was a justifiable experiment."[78]

Singer holds that his position is not based on speciesism. In a public challenge he asked if Aziz would be willing to perform the experiments "on human beings at a similar mental level—say, those born with irreversible brain damage."[79] When I asked Singer to clarify his views further, he wrote to me that in his opinion the "similar mental level" would, at least in what he considered to be "the relevant respects," be humans with the mental level of about a one-year-old. He also wrote that he would only be willing to use those humans with the consent of their parents, or if they had been abandoned by their parents. He would not have that criterion for the monkeys since "the human parents will suffer more and longer from the forced removal of their child than the monkey parents."[80]

Many of us would consider highly speciesist the suggestion that a full-grown rhesus monkey, who could head up a troop that would survive in the jungle, and fend for her own young, and hunt or forage, and build housing, is mentally equivalent to any human age one, "in the relevant respects." As for the suggestion that the monkey parents won't suffer so much from the removal of their offspring, it reminds one of dark times in American history when similar comments were made about human slaves. Though their mothers sobbed as their children were sold away, some people commented that though Africans missed their children, it was not in the same way that white women did. And centuries ago historians wrote that some of the Native American tribes "do not love their children" in the same way others do.[81] I was even amused to find, on the Internet, a document that shared the testimony of a man regarding his inter-action, at Nellis Air Force Base, with extraterrestrials known as the Tall Whites. He explained that when the extraterrestrials abduct human children, the acts are justified with "conde-scending comments by Tall White females over how humans don't love their children like the Tall Whites do."[82]

The suggestion that other primates don't mind being parted from their young as much as human primates do seems to me to be a rationalization. It flies in the face of all those hideous maternal deprivation studies, which found the effects of separation on other primates to be much the same as on humans—including the effect on the mothers who had to be held down while their babies were removed. It contradicts what we know about the wild capture of baby monkeys—that their mothers must be shot, as no primate mother will run away from hunters and leave her baby behind. And it even belies what we know of other animals—the cows who bellow for days or weeks after their calves are taken, and the female foxes caught in traps who chew off their own legs to get back to their young. Humans differ from other animals in many ways, but we are increasingly recognizing our emotional similarity—the realm of maternal attachment being one of the clearest examples. Sometimes mothers of various species have been known to abandon their young. And some human mothers will hand over a baby for a vial of crack. But those are aberrations. Mothers of many species, just like mothers of all human races, generally show profound attachment to their young.

Fox and Felix

Justifications for Aziz's work, if we would not accept that work on humans ("normal" or not), must be based on speciesism.[83] Does that mean the work is wrong? I argued earlier that speciesism is a natural instinct. All things being equal, almost anybody, including many animal rights activists, would choose to save a human over another animal. Generally, when we use animals, that is not the choice being made. Sometimes, however, the choice is a little blurry, and I think that is the case with those experiments aimed at alleviating Parkinson's disease.

I grappled with that issue recently as I watched an interview on the Bravo TV series *Inside the Actors Studio*.[84] Suddenly Parkinson's disease had a face other than Felix's—it had the sweetly attractive and popular face of the actor Michael J. Fox. During the interview Fox was warm and engaging. At one point he started to get the shakes and excused himself for a few minutes to take a pill and let it kick in. For the rest of the interview one would not have known he was afflicted, though he told us that there are times when his symptoms are more severe. He seems to cope as well as possible with his disease, and said that he appreciates the chance it has given him to use his celebrity status to reach out and help other sufferers.

Toward the end of the interview Fox mentioned that his foundation had raised $50 million for Parkinson's research. The audience clapped. I thought of Felix, now dead, and of the other monkeys spending the night alone in their cages, with electrodes in their brains, suffering from intentionally induced tremors, while this charming and likable actor spent his night on a New York stage in front of admirers. As I watched Michael J. Fox, with the shakes, I had compassion for his plight and wouldn't wish it on anybody. But my anybody includes Felix and his brethren. All things being equal, of course I would choose to save the life of Michael J. Fox over Felix. But I cannot be sure it is right to make Felix and other primates suffer with Parkinson's-like symptoms, and to live alone in cages and be experimented on, to go through what the activist Mel Broughton has justly described as "a living hell,"[85] and to ultimately be killed, so that either this sweet Hollywood movie star or less glamorous humans with Parkinson's disease will suffer less.

It is reasonable to suspect that my own health could influence my take on the matter. So I look to my friend Lawrence Carter Long, who is the director of advocacy for the Disability Network of New York City. When he was young, Lawrence was literally the poster boy for Cerebral Palsy research. Pictures of his adorable face and awkward body adorned research fund-raising posters. Now Lawrence speaks articulately and eloquently against animal testing, his arguments based both on its dubious morality and on vivisection's questionable aid to humans when compared to other techniques. You might enjoy listening to a KPFK radio interview I did with Lawrence, Dr. Ray Greek, and Kevin Jonas (from Stop Huntingdon Animal Cruelty), now available online at the Watchdog site, www.WatchdogRadio.com. Lawrence's stance weakens arguments proposing that opposition to animal testing would melt under the heat of a serious affliction. While I respect Michael J. Fox's stance, I do believe that even if I were in his or Lawrence Carter Long's difficult position, my stance would be more like Lawrence's.

Still, I noted above that all things being equal I would choose Michael J. Fox's, or, for that matter, Lawrence's or my own life, above the life of Felix. But things are so grossly unequal. Felix is treated like a thing whose life is only worthwhile as a research tool. Must that be so?

Now what if Felix had been rewarded rather than killed at the end of the experiment, as a human might be, or at least should be? I will discuss that idea further shortly. It would bring us a lot closer to "all things being equal" and would make me considerably more comfortable with the testing situation. I would also be somewhat more comfortable with the situation if I had more reason to believe that the use of other primates had been, and still is, necessary to develop the techniques. In Wishart's documentary Aziz claims, "I was one of a group internationally that has shown that a part of the brain that was never associated with Parkinsonism, the subthalamic nucleus, is overactive. . . ." Curious about that, I consulted Dr. Ray Greek, who led me to papers revealing that it has been known since the 1920s, from studies on humans, that the subthalamic nucleus is associated with movement disorders and that irregular movements "were noted to occur in patients with lesions of the contralateral subthalamic nucleus."[86] Could Aziz be that much older than he looks? I doubt it is his good vegan diet. Further, Dr. Greek led me to early reports on the use of deep brain stimulation to control Parkinson's symptoms. Those reports noted that the technique was pioneered by Dr. Alim-Louis Benabid on human patients back in the 1980s, and has been used on humans ever since.[87] That information calls into question the claims that led Peter Singer to give his qualified endorsement of Aziz's research.

I understand the thinking of those who are sure that what we did to Felix and are doing to other primates is justified, simply because we are human and they are not, and any risky procedures should be performed on "them." We should not forget, however, that prior experimentation on animals hardly removes risk for humans. In neurosurgery, as in other medical fields that use animal testing, there are examples of positive effects on other animals that have led to misplaced confidence for human procedures—and to disastrous results.[88] Also, while I think it is reasonable to respect a position that holds that all risk should be taken on nonhumans (even if it is not my position), that position should be respected only to the extent that it is held in a reasonable manner. I could not call it reasonable if the interests of the animals are assumed to count for nothing.

Which brings us back to Aziz: A few months after the BBC documentary aired, the poster boy for animal research publicly condoned the use of animals to test cosmetics, a practice already banned in the UK. He said, "People talk about cosmetics being the ultimate evil. But beautifying oneself has been going on since we were cavemen." He commented scornfully that British society "has a humanoid perception of animals that's almost cartoon-like."[89]

I suggested above that I accept the support for, and do not actively campaign against, animal testing that appears to have a chance of saving human life. There is too much stark, dark, indefensible cruelty in the laboratories for me to devote energy to areas that could reasonably be judged gray. In a society, however, that judges the interests of humans to count highest but the interests of other animals to still count for something, shouldn't any tests done on animals be conducted by somebody for whom we have reason to believe the interests of the animals do indeed count for something? That would not include somebody who supports outlawed cosmetic testing. Even without the information Dr. Greek provided about Aziz's seeming misrepresentations of his discoveries, Aziz's cosmetic testing stance would make me skeptical about his word when he says there is no way besides animal tests to get information he wants.

Why would a society that cares about animals let somebody like Aziz anywhere near them? At the very least his bombastic statements about cosmetic testing should mean that every step of his work, from requests to grants, to any hair he hopes to touch on a monkey's head, should be microscopically examined. He has announced to the world that he supports practices that UK society has deemed unethical. Letting him work with animals is like letting somebody who has announced his support for child pornography teach kindergarten.

Karen Dawn

voluntary guinea pigs

It is accepted that tests on humans would be more accurate than tests on other animals. But we don't test on humans because society feels it would be ethically unacceptable. In some cases one could argue it would be ethically more acceptable. Take those Tasered pigs, for example, who endured horrible pain in experiments that "cannot be done on people." Tasering is so horrible you probably couldn't even find an Internet chat room devoted to people who get off on it. Yet pain that we understand, or that we have even chosen, does not carry with it the suffering of pain to which others have subjected us. While none of us like the dentist, there is nothing sorrowful or immoral in what we experience in the dentist's chair. And despite the extreme pain of Tasering, one can imagine there could be brave military personnel who would willingly put themselves through a Taser experiment in return for a large monetary compensation—or an early retirement or pension for their family. If there were real risk that the experiment could lead to death, then it would appropriately be performed (only voluntarily) in return for a big payoff, on those who had little life to lose. That might be the terminally ill, or perhaps prisoners who wish to transfer death sentences to life sentences. The very idea of Tasering prisoners or anybody else makes us uncomfortable—it seems like a return to the barbaric days of corporal punishment. But is Tasering utterly innocent animals civilized?

The idea that it is not ethical to use human volunteers in experiments but is ethical to do absolutely anything we want to nonhumans should be questioned. Occasionally the reverse will be true. I am not suggesting, of course, that we do lethal dose toxicology tests on people—or on anybody else. Those barbaric relics of the World War I era need to become a thing of the past.

Louise Wilson/Getty Images

"If you agree with vivisection, go and be vivisected upon yourself" (**Morrissey**).

Nic Fleming, "Medical Tests on Great Apes Should Not Be Banned, Says Research Chief," The *Daily Telegraph*, June 3, 2006.

Thanking the Monkey

Tests that reasonable and compassionate people might deem necessary, but for which humans will not volunteer, could be conducted in such a way that even those generally opposed to animal testing might possibly consider them ethical. Because we don't speak their languages, and vice versa, we cannot ask other animals how they feel about a situation—but we can make some educated guesses.

Knowing that other animals think, and knowing their emotions are similar to ours (see "Are Rats and Mice Animals?" discussed earlier in this chapter), what we can do is live by the golden rule: We can do unto them what we would be willing to do unto ourselves. That could actually allow for some animal testing.

Take Felix, for example: Aziz suggested that Felix's testing would be unpleasant but not painful. Activists from the SPEAK campaign said otherwise, stating on their Web site: "We can now disclose that Felix had the top of his skull sliced off, a procedure that has been documented through human and nonhuman primate research as extremely painful. Electrodes were forced into his brain and then he was fitted with a cranial chamber."[90] (I know nothing of the anesthesia practices in Aziz's lab so won't comment on the pain of the actual operation; in some laboratories anesthesia is generally adequate and in others it is not.) Nobody thinks poor little Felix was having any fun. But what is also apparently not in question is that the tests were not deadly, or even permanently debilitating. I can imagine that some people would be willing to submit themselves to those tests in return for a luxurious retirement. Since tests on humans would be by far the most accurate, we should consider using human volunteers who have something to gain (not powerless people in developing countries, as in *The Constant Gardener*!). But if we use nonhumans, why not treat them in the way we would want to be treated? Some of us would consider a cushy retirement to be a fair trade for a few grueling months of discomfort in which we were helping others in need. So why not reward animals similarly for their service? If, after the experiment, Felix had been returned to his mother and family, and they had been sent to live in the loveliest sanctuary imaginable, might that not be exactly the way we would have wanted to be treated and might the whole scenario—including Sean's restored ability to walk—be acceptable?

It would be expensive—but given some of the things our society spends money on, rewarding those we use in medical research could be deemed a worthy expenditure. And there is a real upside to the expense: If those who profit financially from animal torture, such as the drug companies, were expected to fund beautiful sanctuaries to which every animal used was retired after a single study, then we could assume that animals really would be used only when there were absolutely no alternatives. Under such circumstances, we could predict a vehement acceleration in research into alternatives, and predict that animal research would dry up as profitable alternatives were discovered.

The Unnatural History of Chimpanzees

Chimpanzees: An Unnatural History is a beautiful Emmy-winning documentary by Allison Argo that aired as part of PBS's *Nature* series.[91] It teaches us about the fates of chimps used by human society. After space exploration missions, for example, the human astronauts got ticker-tape parades, while the chimps, as thanks for their service to our country, were strapped into speeding cars and smashed into walls—to test seat belts. Or they were sold to biomedical research laboratories. As noted in chapter 3, chimps from the entertainment and pet industries are also dumped into laboratories.

Because chimpanzees are now members of an endangered species, they are not killed if laboratory experiments do not kill them. The documentary looks at the sanctuaries that have been opening up to care for them. It introduces us to a number of chimps who were used in research and are now at the sanctuaries.

Ron and Thoto lived for many years at the Coulston Foundation in New Mexico, where chimpanzees were kept alone in concrete cells. The documentary tells us that in the United States it is legal to house chimpanzees alone in cages five feet by five feet by seven feet.

Ron had spent most of his life at NYU's Laboratory for Experimental Medicine and Surgery in Primates (LEMSIP) medical research facility and was moved to Coulston in 1996. For some of the studies in which he was used, he was knocked down—that is, anesthetized with a dart gun—every day for a month. In 1999, he was used in a spinal experiment in which researchers, attempting to find spinal disc substitutes, removed his healthy spinal discs and replaced them with a device. The device was later removed, leaving Ron without part of his spine. We learn that to accommodate his pain from the spinal experiment he was given ibuprofen for three days—as if he had a headache or a case of menstrual cramps.

Thoto was born in Africa, captured at a young age, sold to the circus, then sold as a pet, and then finally sold for research. He is a gentle being who has no teeth—they were pulled out during his days in the circus.

The documentary follows their journey to a Florida sanctuary and their release onto the grassy islands. When the door to Ron's indoor room on the island is opened, he ventures out a little, but on his first day he will not leave the cement area near his door. He has spent his life in a steel cage, most of it in solitary confinement, and he has never seen grass before. He looks confused and goes back into his room. We learn from the PBS Web site, however, that eventually Ron learns to love the outdoors. (You can watch Ron and Thoto's story online at www.pbs.org/wnet/nature/chimpanzees/video3.html.)

When the door to Thoto's cage is opened, he charges straight out. As we watch Thoto explore the island he will now share with Ron and others, the filmmaker's voice-over tells us:

> No one could get Thoto to come in that evening. He chose to spend the night beneath the moon.

The documentary also takes us to the Fauna Sanctuary, outside Montréal, Canada, run by Gloria Grow and her veterinarian husband, Richard. They took in fifteen chimpanzees from New York's LEMSIP laboratory, eight of whom were infected with HIV.

As we meet the chimps it becomes obvious that many had been pets or were used in entertainment before they were sent to laboratories. We learn that Jane Goodall has visited the sanctuary, and she and Gloria agree that it is important to try to provide the chimps with things that could remind them of some time in their past that might have been good. So we see a female putting on beads as Gloria says, "She didn't learn that in the lab." We watch one chimp eating spaghetti with a fork, and another eating ice cream on a cone, then wiping his face with a napkin. It is odd to see these behaviors, and we would naturally prefer to see the chimps swinging joyfully through jungles with their families. But watching them behave like us is touching, and brings home the horror of their having been used for experimentation.

Gloria did her best to keep the chimps comfortable and happy in spacious cages while she fought with the city to get permits to build them islands. In an interview on the PBS Web site, after those permits finally came through, Grow says that "the islands are a place with no bars over their heads." She says, "They can come out, surrounded by water, and look up at the sky without any obstruction. There's a vegetable garden on one of the islands. One of the females, Pepper, likes to pick her own veggies. She'll take her blanket with her and camp out on the island at night. She likes the quiet, away from the rest of the group. Just a chance to go out onto their islands has changed their personalities a lot." Gloria says the chimps even love the islands in winter, and that it's fun to watch them smashing down snowmen.

The Fauna Foundation

Tom

While he is still locked mostly indoors, we meet Tom, the oldest chimp at the sanctuary. He spent about thirty years in laboratories. There isn't a shred of information about his life before the lab, though we are told it is likely that he was captured in the wild. Gloria says of Tom, "It is pretty obvious by some of the things he does here that he was with humans for a certain part of his life. He likes to put socks on his feet. He'll put a jacket on. He puts baseball caps on, and he is obsessed about having them on the proper way. He knows how to use utensils." That's more than I can say for some of my dates.

Tom has a special relationship with Pat Ring, who sold the farm to Fauna and comes back regularly to visit. Pat says he used to be a redneck type but that the chimps changed him. Tom is HIV infected, and Pat says, "I wanted no part of the HIV-infected chimps. I was scared of mosquito bites—I thought we were going to all end up with AIDS. Now there is nothing I wouldn't do for Tommy." We see him kissing the adult male chimp through the bars as he explains there is no way he could get AIDS, except through blood-to-blood contact, and that precautions are taken during medical procedures.

Chimps usually have to be anesthetized for any type of procedure, but Tom has an old foot injury that we see him let Pat treat. Pat says, "We just kind of hit it off and we've been best buds ever since. He loves me and I love him and I can't even tell you the reason—I have no idea."

Pat and Tom, sharing a little quiet friendship.

In a climactic scene, we get to see Tom released onto his island. When the door to his cage is opened the filmmaker narrates, "What happens next is beyond what anyone could have imagined."

Tom runs onto the island as Gloria runs alongside cheering him on from just the other side of a little moat. Then Tom hops onto a huge tree and scales it, as everybody cheers. Gloria shouts, "I think that would confirm that's a wild-caught chimp!"

As the documentary ends, Argo says,

If we could retrace our steps perhaps we would choose to rewrite our history with our closest relative. We can't undo the past, of course, but we can reconsider the future and the cost to the chimpanzee. Thousands like Tom have sacrificed everything so that we might live a little longer or laugh a little louder.

We see shots of chimps in laboratories and in circuses. Then, as the camera moves back to Tom, who spent thirty years in a steel cage and is now at the very top of his tree, Argo says:

> Far from the forests of equatorial Africa, this old chimp can finally survey the strange landscape that has become his home. His trials have come to an end but his story will live on—a reminder of the thousands like him, who are still waiting for a second chance.

Tom in his favorite tree.

The Chimp Act

So that all chimps in research could eventually be retired, in 2000 the U.S. Congress passed the Chimpanzee Health Improvement, Maintenance and Protection Act, known as the Chimp Act. It stipulates that government-owned chimps no longer used in research will be provided sanctuary and receive government funding. As sanctuaries are both cheaper and more humane than research laboratories, the act had nearly unanimous support—until a distressing amendment was added that would allow the retired chimps to be pulled back into research if needed. Some sanctuary groups pulled support for the act. Other groups, and leading primatologists, such as Jane Goodall, supported the act even with the amendment, feeling that it was still a step in the right direction. The criteria that would allow for a chimp's removal would not be easy for researchers to meet, and Goodall felt that animal advocates should take what we could get and then fight tooth and nail if any attempts were made to return retirees to research.

In December 2007, a breakthrough: President Bush signed the "Chimp Haven Is Home" Act, prohibiting the removal of, and research on, retired chimpanzees living in Federal Sanctuaries. I still look forward to the day when the chimps are in wonderful care, however, funded by the companies that profited from their pain.

Are Some Animals More Equal Than Others?

It is heartbreaking to read of Ron, Thoto, and Tom spending decades in research laboratories. That is why I would suggest that if any experiment is deemed absolutely necessary for medical progress, participation in one such experiment should earn an animal a luxurious retirement. The Chimp Act is a step in the right direction. Chimps are an understandable place to start as they are endangered, and because people tend to care most about chimps—probably because they are most like us. But all animals matter.

Some people might argue that insisting that drug companies comfortably retire animals would cut all financial incentives for animal research and would thereby endanger the human health benefits it brings us. I hope I have made the case that financial incentives have led to those benefits being overstated. If we took away the financial incentive for animal research without abolishing it, and put that incentive into other research, such as human in vitro and clinical studies, we would see progress that is not dependent on animal research. Many of us who are fans of the free market are also concerned with compassion. In a humane world, torture and the taking of life must be removed from the concerns of commerce.

In 2007, The U.S. National Institutes of Health announced that it will permanently stop new breeding of government-owned chimpanzees for research. The NIH's National Center for Research Resources cited financial reasons for its decision, citing in a statement on its Web site the "fiduciary responsibilities" to maintain the health and well-being of chimpanzees already in its care. It noted that chimpanzees can live at least fifty years in captivity, and that high-quality care for a single animal over its lifespan can cost up to $500,000, and it stated "Therefore, after careful review of existing chimpanzee resources, NCRR has determined that it does not have the financial resources to support the breeding of chimpanzees that are owned or supported by NCRR."[92]

Does that leave much doubt as to the acceleration we would see in the search for animal testing alternatives if we insisted, whenever possible, on comfortable retirement rather than extermination for all animals used in research?

so where do we go from here?

Those working to phase out animal testing talk about the three Rs—Reduction, Refinement, and Replacement:

- *Reducing the number of animals*
- *Refining experiments to minimize suffering*
- *Replacing animals with alternatives when possible*

The Seduction of Reduction

Thanks to public pressure, scientists are devising ways to analyze results so that they can use data from fewer animals.

Encouraging companies (forcing them, if necessary) to share experimental data will greatly reduce the number of animals used.

Changes in FDA drug approval would also have a huge impact. We have noted that drugs are just as effective, though not so profitable for drug companies, once their patents have expired and we can use the generic versions. If the FDA made approval of new drugs contingent on significant improvements over older drugs already on the market, very few new drugs could meet that standard. The amount of money people spend on prescription drugs would plummet along with the number of animals used in bringing new drugs to market. Bad for the drug companies—good for everybody else, especially the animals and those who wish to live in a humane world.

The Gentle Touch of Refinement

I hope the examples given above will cause people to laugh the next time they are told that laboratory suffering is already being kept to a minimum.

Every moment of life for an animal in a U.S. laboratory, life in a cage, is suffering. For a start, minimization of suffering must include an overhaul of housing requirements so that animals may live in spacious environments, with stimulation, and wherever possible in family groups.

Don't forget, the acute suffering during tests on the vast majority of animals, mice and rats, is currently undocumented and federally unregulated. That has to change—all animals must be included under the Animal Welfare Act for the idea of refinement to hold the slightest bit of weight.

Ultimately—Replacement

Advances in the field of alternatives are encouraging. We desperately need, in the United States, laws like those in the European Union that outlaw animal tests where alternatives exist. Instead, our system encourages animal testing even when it is not useful. For their article in *Scientific American* about testing alternatives, Hartung and Goldberg spoke with representatives of nine multinational companies and learned "that all the firms use petri dish or non-mammalian tests, usually involving fish or worms, to decide if a chemical is safe enough to produce. Only then do they perform the life-span feeding studies—to satisfy the companies' lawyers and regulatory agencies."[93]

Government funding for research should reflect the sympathies of a nation that cares about animals and only wants them used when no alternatives exist—in other words, our government should be funding the search for alternatives at a much greater level than it is funding animal experimentation.

Further, our government should stop funding things that make us sick! After thirty years of fighting cancer with animal testing, the number of deaths from cancer has increased. We know that many studies published in scientific journals have linked Western diets to a host of diseases, including cancer. Just think—if animal testing funds were diverted to programs encouraging dietary change, we would finally see disease rates plummet—and by ethical means. Instead, governments support the inhumane factory-farming industry, encouraging Westerners to consume, cheaply, far more meat than is healthy. Then governments support the animal-testing industry to combat the diseases they encourage.

For the three Rs to have any real meaning, experiments that are far from crucial, just a few of which have been outlined in this chapter, must end. There is no room in an ethical society for animal tests on recreational drugs like ecstasy or nicotine, or to "cure" homosexuality or premenstrual tension, or to give us new cosmetics or oven cleaners.

The late Cleveland Amory, a renowned author and animal advocate who founded the Fund for Animals and the Black Beauty Ranch sanctuary in Texas, once took part in a debate with a leading scientist. Amory described various hideous experiments to the scientist, and the scientist defended them. With just a minute or two left, Amory asked about the experiment in which the eyes of a dog were put into a cat and the cat's eyes were put in the dog. The scientist started to justify the experiment. As the debate ended, Amory said, "I want you to know I just made that experiment up, to prove you would defend absolutely anything being done to animals!"[94]

Some of the experiments described in this chapter suggest that Amory's point is shockingly close to the truth. I think we can agree that our society is galaxies away from treating animal testing as a necessary evil. When we seriously start to treat it as such, we are likely to find it is not necessary at all.

chapter seven

THE GREENIES

The earth is common ground

Environmentalists or conservationists fight to save the environment so that future generations of humans can enjoy it. They save countless animals who would otherwise die as their habitats are destroyed.

Back in 2001, I had the pleasure of hearing PETA president Ingrid Newkirk give a talk at New York's Jivamukti yoga center. It was during the period when George W. Bush's administration was attempting to open up the Arctic Refuge for oil drilling. Newkirk said,

> They describe the refuge as uninhabited. But in fact it is inhabited by millions of families. They just aren't human families.[1]

Environmentalists and animal rights activists lobbied hard against the drilling.

Environmentalists also fight specifically for animals—some animals. Where they differ from animal rights activists is in their focus on animals from threatened or endangered species. To an animal dying slowly of a gunshot wound, whether or not she is a member of an endangered species is meaningless. To animal rights activists, who are concerned with her suffering, and with her very right to live, whether there are millions more of her kind is irrelevant—just as the taking of any human life is not inconsequential simply because there are billions more on the planet.

My lawyer finally got me on the endangered species list.

Fighting Factory-Farm Filth

One issue that is top priority for animal rights activists and also for environmentalists is factory farming. In chapter 5 we saw what it does to the animals who live under its punitive conditions. It is also destroying the earth.

Pig Shit Lagoons and Rivers

Rolling Stone ran a terrific article, by the reporter Jeff Tietz, on industrial pig farming. It tells us that pig farms produce millions of tons of waste per year. The waste is even more toxic than would be expected, because pigs are pumped full of antibiotics and doused with insecticide to keep them alive and growing under horrendous conditions. Their waste is held in big lagoons before it is sprayed on nearby fields, from which it seeps into waterways.

According to the article, "Studies have shown that lagoons emit hundreds of different volatile gases into the atmosphere, including ammonia, methane, carbon dioxide, and hydrogen sulfide. A single lagoon releases many millions of bacteria into the air per day, some resistant to human antibiotics."[2] Tietz got near to one of the lagoons and took a whiff. He writes that later, just thinking about the smell, he has to hold back the urge to vomit. He has been told it nauseates pilots flying at three thousand feet.

Tietz takes a trip over the farm area with one of the pilots and writes that he sees "several farmers spray their hog shit straight up into the air as a fine mist." He continues, "It looks like a public fountain. Lofted and atomized the shit is blown clear of the company's property. People who breathe the shit-infused air suffer from bronchitis, asthma, heart palpitations, headaches, diarrhea, nosebleeds and brain damage."

People who live downwind from the farms learn to keep their windows closed at all times. Venturing into their yards, they are sometimes overcome by the fumes.

Hog waste lagoons often overflow, transforming nearby counties into what Tietz calls "pig shit bayous." Hog farms have spilled millions of tons of pig waste into rivers. The worst single spill to date—twice the size of the *Exxon Valdez* oil spill—dumped 25.8 million gallons of toxic hog waste into the headwaters of the New River in North Carolina. Every single living creature in the river was killed. The riverbanks were blanketed in dead fish.

We noted in chapter 5 (under "Trawling and 'Trash Animals'") that ecologists have projected "all commercial fish and seafood species will collapse by 2048."[3] Factory farms contribute to that disaster not only with pollution: While we know the oceans are being overfished to feed humans directly and to feed the fish on fish farms, a Worldwatch paper has also reported that "about a third of the total marine fish catch is utilized for fish meal, two-thirds of which goes to chickens, pigs, and other animals."[4]

Dairy Farm Delights

Dairy farms handle their waste much as pig farms do. The *New York Times* editors wrote a searing commentary after the burst of a dairy farm reservoir released three million gallons of cow waste into New York's Black River. The editors drew a bleak picture: "In case you have trouble visualizing it, three million gallons of liquid manure is roughly equivalent to the water in six Olympic-

Photograph by Rick Dove/www.doveimaging.com

size swimming pools."[5] The spill caused a massive fish kill in the river, which feeds into Lake Ontario.

Dairy farms appear to have an even worse effect than pig farms on the atmosphere. A *New York Times* editorial about dairy farm pollution discussed "the eye-stinging, nose-burning smell of cattle congestion in rural California,"[6] and a *Washington Post* front-page story told us that California's San Joaquin Valley, responsible for a fifth of U.S. milk production, rivals Houston for the title of worst air quality. It said that 15 percent of the region's children have asthma.[7] The *Los Angeles Times* reported that dairy cows have overtaken automobiles as the number-one air polluter in parts of California, and that in Fresno, in the center of the nation's dairy industry, one in six children carries an inhaler to school.[8]

In the face of those revelations, how nice to learn, from the *International Herald Tribune*, that Dutch dairy farmers are shifting their operations to the United States because Holland "has more effective environmental restrictions than the U.S."[9]

What If the Shit Don't Stink?

While human waste is intensely processed to minimize environmental damage, the waste from farm animals is not. Tietz's *Rolling Stone* article explains that every pig produces three times more excrement than a human. Smithfield is really full of it—as the word's largest pig producer, the company discharges about twenty-six million tons of waste per year. Tietz writes, "So prodigious is its fecal waste, however, that if the company treated its effluvia as big-city governments do—even if it came close to the same standard—it would lose money." He tells us that fixing the problem completely would bankrupt the company.

In other words, the only way to save the environment would also help the animals: get rid of the factory farms. That's why environmentalists and animal rights activists are on the same side of the factory-farm issue—for now. I write "for now," because Tietz discusses programs designed to turn pig shit into alternative fuel. His article tells us that Smithfield is building a facility in Texas to produce clean-burning bio-diesel fuel.

Raleigh's News & Observer tells us that a subsidiary of Smithfield in North Carolina has enclosed 20 percent of a seven-acre hog waste lagoon. Reporter Wade Rollins writes: "Captured beneath the bulging cover is methane gas, which rises off the pond as manure decomposes. It is a potent greenhouse gas trapping heat in the Earth's atmosphere. But methane can also fuel an incinerator or drive a turbine to produce electricity." The article points out that "covering lagoons to capture methane treats only one problem: methane emissions. It does not address such issues as odor, ammonia emissions and other pollutants that have long been a problem on factory farms." There is not a word about the animals.[10]

Smithfield is still a long way from turning all its shit into clean fuel, but what if it were to succeed? Then, to environmentalists, would factory farming still be an issue? Would the intelligent creatures who live inside, bathed in the fumes of their excrement, still matter if that excrement eventually became clean fuel?

A Gory Gorey Truth

Many of us who saw *An Inconvenient Truth*[11] were stunned that the film ended with tips for lifestyle changes to help reduce global warming but failed to recommend eating less meat. I recall that in the film Al Gore said that it is hard to get somebody to see something when his livelihood depends on his not seeing it. I think of the scenes of the Gore family's Angus ranch and wonder if, despite his dedication to the environmental cause, his family's cattle-ranching background has made it impossible for Al Gore to make the connection between meat and global warming.

Others have made the connection. In late November 2006, the United Nations Food and Agriculture Organization published a report titled "Livestock's Long Shadow."[12] It informed us that the livestock sector generates, at 18 percent, more greenhouse gas emissions as measured in CO_2 equivalent than that generated by transportation. It said that the livestock sector emits, mostly from manure, 65 percent of human-related nitrous oxide, which has 296 times the Global Warming Potential of CO_2.

The report also discussed cow contributions to methane in the atmosphere: Livestock accounts for 37 percent of all human-induced methane, and methane has twenty-three times the warming power of CO_2. If you work in a building with a burrito joint next door, and the elevators are crowded after lunch, you probably already know that.

The *New Scientist* has reported on a study done at the Institute of Livestock and Grassland Science in Tsukuba, Japan, which showed that "producing a kilogram of beef leads to the emission of greenhouse gases with a warming potential equivalent to 36.4 kilograms of carbon dioxide" and "is responsible for the equivalent of the amount of CO_2 emitted by the average European car every 250 kilometers, and burns enough energy to light a 100-watt bulb for nearly 20 days."[13]

While Gore's *An Inconvenient Truth* counted the destruction of the Amazon rain forest as a significant contributing factor to global warming, it avoided reference to the cause of that destruction. The UN report, however, noted that 70 percent of former forests in the Amazon have been turned over to cattle grazing.

Mutts © Patrick McDonnell/King Features Syndicate

Oddly, the UN report makes various recommendations about the need to treat manure more effectively and change the diets of cows to cut down on methane, yet cutting down human consumption of meat and dairy products is not even mentioned as an option.

The *New York Times* editorial on the UN report ended with,

> The human passion for meat is certainly not about to end anytime soon. As "Livestock's Long Shadow" makes clear, our health and the health of the planet depend on pushing livestock production in more sustainable directions.[14]

Not all of us are willing to swallow that paper's conviction about the human passion for meat. There is a new generation that is passionate about the planet—and many of them will gracefully adjust their eating habits in order to save it.

Little Piggies—Eating Everybody Else's Food

Some environmentalists, noting the connection between meat eating and our impending environmental disaster, are willing to recommend behavioral change. In his book *The Future of Life*, Edward O. Wilson points out that meat consumption not only pollutes the environment and contributes to global warming, but also taxes resources more than the consumption of other foods. He asks, "Stretched to the limit of its capacity, how many people can the planet support?"[15] Then he writes:

> The current world production of grains, which provide most of humanity's calories, is about 2 billion tons annually. That is enough, in theory, to feed ten billion East Indians, who eat primarily grains and very little meat by western standards. But the same amount can support only about 2.5 billion Americans, who convert a large part of their grains into livestock and poultry.

Grain fed to the animals eaten by omnivores is consumed by those animals in much greater amounts than when it nourishes people directly. And, of course, to grow the four times as much grain it takes to feed an American meat eater, it takes four times as much water.

Wilson's estimate is based on the assumption that Americans eat a mixture of animal products. But some of those animal products are more energy-intensive than others. An article on Green Living in *Newsweek* reports, "In fact, it takes about 16 pounds of grain to produce one pound of beef."[16]

That *Newsweek* article at least mentions the need for behavioral change. It quotes the president of the Union of Concerned Scientists:

> You don't have to be a vegetarian—just take a break once or twice a week. If everyone tried to do something that simple, it could have a huge environmental effect.

Tim Mosenfeider/Getty Images

"Millions of people are learning that a vegetarian diet is the healthy choice for themselves, the earth, and the animals" (vegan songstress **Fiona Apple**).

PETA PSA quoted in the *Washington Post*, November 22, 1997, p. B3.

Here's Fiona hanging out at the PETA office.

Courtesy of Fiona Apple and PETA

Karen Dawn

Indeed, any change helps. Wilson, however, was willing to write, "If everyone agreed to become vegetarian, leaving little or nothing for livestock, the present 1.4 billion hectares (3.5 billion acres) of arable land would support about 10 billion people."

A note: Some have argued that vegetarians are responsible for more animal deaths than meat eaters because of all the animals who die during the farming of crops—they are run over by tractors, have their burrows destroyed by plows, or are killed by insecticides.[17] That argument, however, ignores the far greater number of crops one must harvest to feed the animals eaten by an omnivore than to feed the vegetarian directly. Plant-based diets therefore save not only those animals who would otherwise be eaten, but also those living in the extra fields of grain harvested to feed the animals on which the omnivore feeds. As noted above, that overall lower consumption also means that plant-based diets have the capacity to save fellow humans starving for lack of grain.

Even if ruminant animals are raised, such as grass-fed cattle, the destruction is far greater for an omnivorous diet. Cattle grazing, like crop harvesting, still kills some animals of the field, and many more fields are needed to feed a person with an omnivorous diet than a person eating a vegan diet. The researcher Gaverick Matheny notes,

> In one year, 1,000 kilograms of protein can be produced on as few as 1.0 hectare planted with soy and corn, 2.6 hectares used as pasture for grass-fed dairy cows, or 10 hectares used as pasture for grass-fed beef cattle.[18]

Environmental groups have tended to tiptoe around the fact that plant-based diets save the environment. It is heartening that some environmentalists are beginning to utter the previously unspeakable gory Gorey truth, and, like animal rights groups, to call for significant dietary change.

In the liner notes for his platinum-selling album *Play*, **Moby** tells us he is vegan for various reasons, including his love of animals, and also because:

You can feed lots more people with grain directly than by feeding that grain to a cow and then killing the cow. In a world where people are starving it seems criminal to fatten up cows with grain that could be keeping people alive.

The raising of farm animals is environmentally disastrous. All of the waste from animal farming gets washed into our water supply, poisoning our drinking water and fouling our lakes, streams, and oceans.

Liner notes by Moby, from his album *Play*, 1999 V2 (label). Reprinted with permission.

Photographs compliments of Moby

Karen Dawn

war on the whales

Deadly Decibels

Environmentalists and animal rights activists often join forces against attacks on marine mammals.

In attempts to detect potential terror threats, the U.S. Navy conducts sonar tests. Evidence has mounted for years, and the navy now acknowledges, that sonar tests kill whales. For example, federal marine specialists concluded that navy sonar was the most likely cause of the stranding of two hundred melon-headed whales in Hawaii while a major American-Japanese sonar training exercise was taking place at the nearby Pacific Missile Range Facility.[19]

Sonar waves sweep across tens or even hundreds of miles of ocean, revealing objects in their path. Some sonar systems can generate 235 decibels under the water—as loud as a space shuttle launch.[20] By the navy's own estimates, low-frequency active sonar waves can retain an intensity of 140 decibels even three hundred miles from the source. That's a hundred times more intense than the noise aversion threshold for gray whales.[21]

Whales in areas where tests take place are found stranded on beaches with blood coming out of their ears. Animals that came ashore during one mass stranding had developed bubbles in their organ tissue that looked like "the bends."[22] The bends kills divers who surface too quickly from deep water—exactly what you'd expect a whale to do when the water is blasted with hundreds of decibels of sound—or maybe even just 50 decibels if it's 80's music.

In 2003, environmentalists led by the Natural Resources Defense Council (NRDC) won a major victory when a federal court ruled illegal the navy's plan to deploy LFA sonar through 75 percent of the world's oceans. None of the limits apply during war or heightened threat conditions. The Bush administration nevertheless pushed legislation through Congress that exempts the U.S. military from core provisions of the Marine Mammal Protection Act. That put the ruling limiting deployment of LFA sonar in question. In July 2006, the Ninth U.S. Circuit Court of Appeals dismissed the Bush administration's effort to overturn the 2003 lower court decision limiting the peacetime use of low-frequency sonar.[23] While that was good news, the court battles about the use of sonar, both low- and mid-frequency, are ongoing, and the whales are in jeopardy.

Few deny that the navy needs to be able to test sonar, but how many of us think the human species has the right to blanket the ocean with it? Can't we at least expect the navy to choose testing sites that avoid whale-feeding and whale-breeding areas and migration routes? The NRDC suggests that care must be taken to listen with passive sonar before switching on active sonar, to ensure that marine mammals are not in the testing area. A *New York Times* editorial noted that the navy had chosen sites for practical reasons, such as proximity to ports and air stations, and suggested, "Whales were at best an afterthought, but they deserve to be a priority."[24]

I must share a letter to the editor published in the *Seattle Times* after the navy announced a plan to send dozens of dolphins and sea lions, who had been trained to detect and apprehend waterborne attackers, to patrol a military base in Washington State:[25]

Dear Secretary of the Navy:

The International Marine Mammal Guild hereby acknowledges the receipt of your request for assistance in protecting your group of humans, your multiheaded nuclear missiles housed in your nuclear submarines, and your homeland from harm ["Navy wants dolphins, sea lions to defend us," page one, Feb. 13].

Please review our many requests for your assistance in protecting us and our homewater from human-caused pollution, habitat loss, food scarcity, ocean sonar testing, Japanese "scientists" and Icelandic harpoons. After doing so, you may contact our office about setting up a meeting.

We look forward to working out a mutually beneficial agreement in a timely fashion.

Best regards,

The International Marine Mammal Guild Interspecies Subcommittee for Human Affairs, submitted by our human contact,

Bruce Clifton, Seattle[26]

The Warring Rainbow Warriors

Environmental groups also protect marine mammals, such as whales, dolphins, and seals, from hunters. Greenpeace put marine mammal protection on the map. But Greenpeace is not without its detractors. Paul Watson, who now heads up the marine mammal protection group Sea Shepherd, was asked to resign from the Greenpeace board of directors when they accused him of having broken the principle of refraining from violence during his opposition to a seal hunt. As he tells it, he was accused of having assaulted a sealer and destroying his property:

> "You took a seal club and threw it in the water."
> Watson's response:
> "Well holy shit. The jerk was about to bash in the brains of a baby seal, for Christ's sake. I was not going to stand there and watch it."[27]

The difference in approach continues today as both Sea Shepherd and Greenpeace take to the seas alongside Japanese whalers. Greenpeace "bears witness" and Sea Shepherd deploys prop foulers—long strands of cable that catch and tangle a whaling vessel's propeller. Sea Shepherd also sinks whaling vessels. The vessels have all been in port with nobody on board. Sea Shepherd asks for Greenpeace's cooperation but Greenpeace dissociates itself from Sea Shepherd's efforts. But Watson surely has not encouraged the cooperative spirit, as he refers to Greenpeace as "the Avon Ladies of the Environmental Movement"—partly a reference to the armies of fundraisers who go door to door.

A gripping *National Geographic* article about Sea Shepherd's attack on Japanese whaling vessels crystallized the differences in approach.[28] Greenpeace followed the whalers and filmed them—with no immediate impact on the killing. (Watson accuses Greenpeace of spending decades making "snuff films.") Then Sea Shepherd showed up, attempted to disable the boats, and played dangerous games of chicken on the high seas. Sea Shepherd shut down the whaling for days at a time—clearly more effective than Greenpeace in the short term. But the risks taken by the Greenpeace crew, the effect

At the 2007 Sea Shepherd Breaking the Ice benefit, Captain Paul Watson presented the first Sea Shepherd "Rock the Boat" Award to Anthony Kiedis, Flea, and the Red Hot Chili Peppers for their years of support.

Here's Anthony hanging out with Captain Paul and Sea Shepherd director's board member Susan Weingartner

Photographs by Denise Borchert

of their "snuff films" on public opinion, and the psychological effect that Greenpeace's constant pursuit surely has on the whalers makes the "Avon Lady" tag, while amusing, seem unfair.

While Greenpeace and others condemn Sea Shepherd's militant tactics, Watson says, "If you can find me one whale that disagrees with what we're doing, we might reconsider."[29]

Back in 1996, Greenpeace had a more widespread falling-out with the humane and environmental movements over its support of the "Dolphin Death Act." Other groups argued the act would have weakened earlier legislation that protected dolphins from fishermen who chase down dolphins and encircle them with nets, drowning many, in order to catch tuna on whom the dolphin feed. The Dolphin Death Act was opposed by the Humane Society of the United States, by most of the leading conservation groups, such as the Sierra Club and Audubon, by dolphin experts, such as Jacques-Yves Cousteau and Jean-Michel Cousteau, and by members of the legislature who had supported the earlier successful bill, such as senators Barbara Boxer and Joseph Biden, and representatives Gerry Studds and George Miller. Its supporters? Greenpeace—and legislators well known for their war on wildlife. It was a confusing time in the history of dolphin protection.

About his departure from Greenpeace, Watson writes, "With Greenpeace I had seen an organization that would grow, prosper, and evolve into a bureaucracy. But that's what it needed to be spread around the world."[30] Clearly Greenpeace is not just a bureaucracy. The group does valuable work. But some worry that its success in the mainstream has come at the expense of some of its value.

Green Meanies

Death by Do-gooders

It is sad to shift from the discussion above of the NRDC's efforts to save the whales to its support of widespread animal poisoning in the testing of chemicals we already know to be toxic.

The Environmental Protection Agency is developing a massive animal-testing program to test the effects of chemicals on the human body's hormones. While PETA has called for the Environmental Protection Agency to increase its funding for and use of nonanimal test methods, the Natural Resources Defense Council has aggressively lobbied for several large animal-testing initiatives. The NRDC even sued the EPA in an attempt to force the agency to begin the testing process sooner.[31]

The World Wildlife Fund held the same position on the EPA's massive testing program but softened it just after receiving a request from Paul McCartney. The group then released a heartening statement that "urges EPA to rely to the fullest extent possible on validated non-animal screens and tests" and says that "WWF believes a number of chemicals have been shown to be sufficiently harmful to warrant imposing bans or severe restrictions on their use now, without need for further tests."[32] While the statements of the group are positive, PETA fears they may be "greenwashes" that are not being backed up with action.[33] We'll know more in time. You can check out www.meangreenies.com to learn more about these issues.

Hunting—The Great Divide

Between animal rights folk and conservationists, hunting is the great divide. Animal rights activists oppose hunting because we are concerned not just for the life of the species but also for the lives of the individuals within it, and for their suffering.

When animals are killed outright, they may not suffer greatly, but their companions and their babies suffer. And far too often they are not killed outright. How often does a hunter make a clean shot to the head or heart? In a paper written for the hunting group Ducks Unlimited, the Game, Fish and Parks' head waterfowl biologist (hardly a biased animal rights activist) reported:

> Numerous U.S. and Canada research studies have been published involving trained observers. . . . These studies document wounding rates of more than 30 percent. Therefore, if you reconcile hunter and

trained observer reports, the wounding rate on ducks is at least 25 percent. Wounding losses on geese are similar. This means that approximately 3.4 to 3.7 million ducks and geese go unretrieved each year in the U.S. and Canada combined.[34]

Millions!

Few of us stop to think what that means. I only really got it while reading a compelling novel by the acclaimed writer Chris Bohjalian titled *Before You Know Kindness*. The story features an animal rights activist whose shoulder is shattered when his twelve-year-old daughter accidentally fires her uncle's rifle. It isn't an animal rights novel; in

SESAME STREET ADDRESSES HUNTING

Bizarro © Dan Piraro/King Features Syndicate

fact, Bohjalian seems to have antipathy for some animal rights activists. But the novel is loaded with compassion for animals. His many lengthy descriptions of the agony of a bullet wound are likely to make his readers think of the animals injured by hunters. Here's an example:

> Spencer knew the button that would propel a burst of morphine down the intravenous drip and into his blood was within inches of his left hand, and he needed to move only half an inch on his back and then stretch his fingers and press. But the very idea of that first motion—sliding over barely the width of his pinkie in his bed—filled him with dread. He feared if he did he would feel again those excruciating spikes of pain that had filled his shut eyes with omegoid bursts of white light, caused his back and shoulder and arms to recall in all too precise detail what it felt like to be pulverized—the bone ground to powder, the muscles and tendons shredded and whisked. He spasmed at the memory, whimpered like a terrified puppy. But he knew he could use some more drugs. As it was, even lying still in the middle of the night, he was in more agony than he had ever endured in his life, and wanted either a rush of morphine pumped into his system now or a nurse to appear in the dim light in his hospital room with one of those tiny paper cups full of meds.[35]

The millions of animals per year who get shot and get away die slowly in the kind of agony described above, in too much pain to forage, and with no nurse to bring them food or morphine.

Most of the well-known conservation and environmental groups officially state that their position on hunting is neutral—they say they neither support nor oppose it.[36] But a closer examination shows close ties to hunting groups and much support for hunting.

Audubowhunt Society?

The Audubon Society holds that hunting is a legitimate means to control populations that threaten ecosystems. One might assume they would support only the most efficient weaponry available. What a shock to learn that when sixty deer were said to be damaging the ecosystem on the Audubon's 285-acre reserve, the society held a bow hunt.[37] Perhaps Audubon didn't want the sound of gunfire to frighten the birds.

Bizarro © Dan Piraro/King Features Syndicate

In a study conducted to find wounding rates during bow hunts, researchers radio-collared eighty deer during a two-year period. Of the twenty-two shot by bow hunters, "11 were recovered by the hunter, resulting in a 50% wounding rate."[38] Bow hunting is for fun, not for clean kills to control populations. People who care about animal suffering should know they might be supporting that kind of fun when they support the Audubon Society.

The Ghost of Mr. Muir

John Muir must be turning in his grave. He founded the Sierra Club in 1892 to protect the untamed wilderness and the animals in it. His many writings show love and respect for all creatures. In a 1904 letter he referred to hunting as "The Murder Business" and to the "rights of animals and their kinship to ourselves."[39] Many of us grew up in families that supported the Sierra Club and John Muir's mission. The organization took a neutral position on hunting for a century. In the 1990s, however, under a new board of directors, the club launched the Hunter and Angler Outreach Campaign. On April 21, 2006, the 168th anniversary of John Muir's birthday, Sea Shepherd's Paul Watson resigned as a national director of the Sierra Club, in protest of the club's sponsoring of a contest entitled "Why I Like to Hunt." The contest offered a hunting trip to Alaska as the first prize. Watson writes, "The Sierra Club founded by John Muir today features pictures of smiling Sierra Club staff posing with their recently slaughtered trophy animals."[40]

National Wilddeath Federation

The National Wildlife Federation (NWF) Web site proudly displays the motto "Inspiring Americans to Protect Wildlife for our Children's Future." As I type this chapter that Web site's front page features polar bears—a widely popular species—and a discussion of global warming. That image plays beautifully to the animal-loving public who send donations. A closer examination of the Web site, however, leads us to the state affiliates of NWF. The affiliates, with a less keen eye on public donations from big-city liberals, are more up-front about their basic goal—to protect hunting. The Washington affiliate, for example, the Washington Wildlife Federation, states on its Web site's front page, "We're an organization of conservation-minded hunters, anglers, and other outdoor enthusiasts who are working for healthy habitats and public access."[41]

Michigan has a one-hundred-year tradition of protecting doves from those who wish to use them for target practice. In 2006, the people voted, in a landslide, against a ballot initiative that would have opened up a hunting season on those gentle symbols of peace. The public voted against the efforts of the National Wildlife Federation's Michigan affiliate.[42]

Bloodbath in the Nursery

Every year Canadian fishermen slaughter hundreds of thousands of baby seals. They head for their nurseries on the ice floes and shoot the babies or, to prevent putting bullet holes in their skins, club them to death. Canadian law allows the pups to be killed after twelve days of age.

Fishermen suggest that the seals are responsible for the depletion of cod. That's a hard argument to swallow given that there have always been seals in the Canadian oceans and plenty of cod—at least until the fishermen, who are also the seal hunters, arrived. In a piece published in Canada's *National Post*, Matthew Scully wrote, "Marine biologists uniformly tell us that commercially fished cod comprise no more than 3 percent of a harp seal's diet. The seals eat squid, skate and other predators of cod, and so, in this and other ways, actually aid the fishermen if given a chance."[43]

The *National Post* came out against the seal slaughter in an editorial that tells us:

> A panel of veterinarians who observed the hunt in 2001 reported that as many as 42% of the seals whose carcasses they studied appeared to have been skinned while they were still conscious. Whereas the preferred manner to kill a seal is to render it unconscious with a single blow and then bleed it to death, live seals are often dragged across the ice by hooks before being skinned. In 40% of filmed cases studied by the same vets, injured animals were left on the ice after being clubbed once before hunters returned to hit them a second time. And that doesn't even include the seals that are shot by hunters but escape under the ice, where they die agonizing deaths.[44]

On *Larry King Live*, the premier of the Canadian provinces of Newfoundland and Labrador noted that the World Wildlife Fund supports the seal hunt.[45] Besieged by public protest, the group called for a letter from the premier "clarifying" the group's position, which is neutral. Parts of the letter are published on the WWF Web site.

So where did the prime minister get the idea that WWF supported the hunt?

In 2003, the World Wildlife Fund Web site included the statement, "WWF is not an animal welfare organization. We support the hunting and consumption of wild animals provided the harvesting does not threaten the long-term survival of wildlife populations. WWF has never opposed a sustainable seal hunt in northern or eastern Canada."[46]

In a keynote address in 2005, the president emeritus of World Wildlife Fund Canada said, "We are also not an animal rights group, a humane society, an animal welfare organization, or one of those groups opposed to hunting, trapping or the seal hunt."[47]

It is understandable that even without the words "We support the seal hunt," a statement supporting hunting in general and specifically announcing no lack of opposition to the seal hunt would be taken as support.

Paul Watson

The prime minister put WWF on the spot, and the current statement (as of 2007) is less vehemently unopposed. Interestingly, it acknowledges the role of public pressure in the subtle shift in stance:

> WWF recognizes that many people around the world feel very strongly about the seal hunt and we understand the sensitivities and concerns around this issue. WWF does not defend hunting of the harp seal, but our specific focus is on conservation of the world's endangered species and their habitats. Because the harp seal population is not a conservation risk at present, the hunt has not been a conservation issue for WWF. We leave work regarding the seal hunt and the humane treatment of animals to specialized organizations which focus exclusively on these issues.[48]

In a few years the statements have changed from "WWF has never opposed a sustainable seal hunt" to "WWF does not defend hunting of the harp seal." Both could be taken to be in line with the group's "neutral" stance, but note the change of emphasis—a welcome change.

Killing to Be Kind?

While the majority of people don't hunt, and don't like to see wildlife blown away for fun, arguments about ecosystem conservation can be persuasive. Also persuasive are suggestions that the animals will die slow, horrible deaths from starvation if we don't hunt them. But is that worse than slow horrible deaths from gunshot wounds?

Moreover, animals won't starve to death if we don't hunt them. When there is not enough food to support her offspring, a female deer will not ovulate and reproduce. Other species show similar adaptation of their reproductive cycles to food availability. For example, a *New York Times* article discussed the ability of red squirrels to adjust their reproduction to the amount of food available.[49]

The Right to Arm Bears?

Bears endanger people and must be controlled—or so say those who have recently reinstated bear hunts in the northeast United States.

In 2003, New Jersey held its first bear hunt in thirty-three years. Bears had become
a nuisance, showing up in people's backyards, getting into the trash, and making people
fear for the safety of their children. A hunt was allowed, even though state officials already had the ability to kill problem bears. Those killed during the hunt were likely just
minding their own business. Shooting bears at random is like trying to reduce crime by
shooting into a crowded room where there might be a felon or two.[50]

Public opinion about the hunt was somewhat mixed until the public was faced with
the death of a cub. The New Jersey *Star-Ledger* reported:

> And in a sad footnote, morning rush-hour commuters on Route 23
> in West Milford got a front-row view of the hunt, when a mortally
> wounded cub staggered out of the woods, lay down with his head
> resting on the road, and died.
>
> "It just broke my heart, sitting there in traffic watching him die,"
> said Kari Casper, a fourth-grade teacher from Vernon who was on
> her way to work in Lincoln Park.
>
> "He was just a little guy and looked so lonely, lying there with
> snow on his paws," said Casper, who said she cried as she watched
> the bear for about 20 minutes as she sat in traffic.
>
> West Milford police said they received at least 25 calls around
> 7:30 a.m. about the injured cub.
>
> "When we got there, he was dead and there were all these cars
> pulled on the side of the road and people crying," police dispatcher
> Lorraine Steins said. "I feel like we should apologize to the bears."
>
> Steins said a hunter showed up some time later, identified the
> cub as his kill and took it away. The hunter was not cited because
> he had the proper state permit and had tracked the wounded
> bear out of the woods. There is no way to control where a wounded
> bear will go to die.[51]

I finally remembered—red with hunter, white with fisherman.

That cub's death saved hundreds of other bears, as it galvanized opposition to the hunt. There was no New Jersey bear hunt the following year. Bradley M. Campbell, the commissioner of the state's Department of Environmental Protection, said, "We simply can't conduct a hunt at that level of controversy."[52]

The next year, however, nearby Maryland held its first bear hunt in fifty-one years, and the following year the New Jersey bear hunt was back. Apparently the controversy had died down, and even though not one person had been seriously injured, let alone killed, by a bear, Campbell said a culling was necessary because "communities are fearful for their safety and the safety of their families."[53] Newspaper reports told us about the protective and oh-so-brave hunters perched in trees, having left pots of honey right underneath to lure the unsuspecting bears.[54]

The big news about Maryland's second bear hunt was that an eight-year-old girl made the first kill. Newspaper readers, perhaps accustomed to seeing little girls photographed with puppies, were shocked to see this one sitting on a rug with the dead bear beside her.[55] The next year there was no bear hunt in Maryland.

The future of bear hunts is up in the air. The hunting lobby is strong, but public support for the hunts is not, and the animal protection movement is gathering strength.

Bambi's Revenge

Ironically, it came out later that the little bear who died by the side of the road in New Jersey had not been shot by a hunter, he had been hit by a car.[56] The hunter had lied in order to claim the bear's body. The car collision is not surprising. The gun lobby points to human deaths from collisions with animals to justify population control hunts, but a study of statistics from thirty-three states showed the number of those collisions is three times higher during hunting season, as animals dash out of the woods to escape hunters.[57] Unfortunately, people who have nothing to do with the hunt are killed in those accidents.

The Removal of the Fittest

The whole idea that hunting controls populations is a myth. The great prize is the huge bear—usually male—or the buck with a glorious rack of horns. But we saw in chapter 2 (under "Females First") that neutering males has far less impact on population than spaying females. Similarly, killing off males does not control population—unless you kill all of them. One male can impregnate many females. If hunters really wanted to cut down populations, they would kill does rather than bucks. Bear hunters would seek out the smallest bears, likely to be young females with many reproductive years ahead.

The hunter's attraction to the biggest and strongest, however, does affect the population—not in number but in health and robustness. In nature we see a real kind of culling, where the elderly, weak, and sick are the most likely to fall to predators, such as wolves. But human hunters, with guns (and pots of honey), need not chase down their prey. Instead of picking off the weak, they take the most pride in killing the strong. So while they don't have any lasting impact on the numbers of animals, they affect the genetic makeup of populations and bring about the deterioration of the species they hunt.

Nature Conservancy Annihilations

One and a half centuries ago ranchers brought pigs to Santa Cruz Island, off the coast of California. Over the years some pigs escaped the farms and bred. The island developed a population of about two thousand wild pigs who trample that island, destroying native plants. Golden eagles prey on their babies, and also on the island fox, an animal now on the endangered species list. The Nature Conservancy, which owns 75 percent of the island, has suggested that it is the baby pigs, rather than the foxes, who actually attract the eagles, and that the pigs thereby cause the deaths of the foxes. I don't follow the logic—if the eagles eat the foxes, why wouldn't some eagles come just for the foxes if the pigs were removed? It is understandable, however, that the Nature Conservancy wants the plant-trampling pigs gone.

One would have hoped the group, funded by those who love nature and animals, would have found a humane method to achieve its goals. But the Nature Conservancy has hired pilots to chase down pigs and shoot them. Clean shots from planes are unlikely. The pigs are being terrorized and dying in agony.

Given how many millions of species of plants and animals have gone extinct due to the preponderance of humans, it is painfully ironic to think humans feel they have the right to conduct mass animal exterminations to protect species of either plants or other animals who happen to live on an island where humans like to see them when they go hiking.

Nobody is suggesting we do nothing to control the pigs. But having erected fences and divided the island into sections to enhance the kill effort, we wish the Nature Conservancy had chosen to relocate the pigs to one or two sections, and dart them with contraceptives. Pig life spans are short. The pigs would have died out completely in about ten years, without extermination. The foxes currently being bred for relocation onto the island could have waited ten years—in another decade there would be that many more of them to introduce back onto the island. The whole process would have been slower and more expensive, but we look forward to the day when fastest and cheapest are not the driving forces when we are dealing with lives. We hope the need for humane solutions will soon be taken as a given, regardless of the expense. Conservation groups should be among the first to adopt that attitude.

Drive-by Population Control?

We should not accept the idea that killing is an appropriate way to deal with overpopulation. One look at Los Angeles freeways at peak hour tells us there are too many people in Los Angeles. From time to time Angelenos take control of the matter and start shooting at each other. But nobody thinks that would be the right government sanctioned policy for dealing with overpopulation. Why is it considered appropriate for animals?

Bizarro © Dan Piraro/King Features

Natural Allies

The environmentalists are not the enemy! We just wish they were the kinds of friends that many people assume they are. In some areas, such as habitat preservation and in the fight against factory farming, they are wonderful friends. In other areas they disappoint us. It is hard to face essays published in their magazines describing hunters as their "Natural Allies."[58]

Perhaps those essays should not be surprising. It is natural for a movement to seek out strong allies, and the hunting lobby, while not large, is strong. (We'll look more closely at that strength in the chapter 8 section "Living with Electile Dysfunction.") As the animal protection lobby gains strength, we may see environmentalists more eager to align with us.

We see hints of their approach. It is heartening to see the World Wildlife Fund, under pressure, temper its call for animal tests, and hedge away from its tacit support of the seal hunt. We look forward to the day when its officially "neutral" stance on all hunting leans away from support. Indeed we hope the officially "neutral" but somewhat supportive stance on hunting, and animal testing, will change for all of the environmental groups as the issue of animal rights becomes part of mainstream thinking.

Meanwhile, those who support environmental causes, and who care about animals, cannot assume their donations to environmentalists or conservationists are animal-friendly. They can, however, make their voices heard within the environmental movement. We are thankful for environmentalists and conservationists—for the protection they give to animal habitats and for their work to preserve all species. We must reach out to them, and urge them to consider not only those species but also the value of the lives of the individuals within them.

SEAL OF THE PRESIDENT OF THE UNITED STATES

Anthony Freda

chapter eight

compassion in action

We can help the animals by making changes in our own lives. Many of us, however, hope to have a wide sphere of influence, persuading many people to adopt more animal-friendly lifestyles and encouraging society to abandon cruel practices. How can those of us who care make the most difference?

The Best Government Money Can Buy

It should be heartening to learn that most people care about animal cruelty. In a Gallup poll, 72 percent of respondents said they agree with the goals of the animal rights movement—30 percent strongly agree.[1] Unfortunately, that general concern for animals does not always translate into behavior. Often people are unaware of the need for change. They assume, wrongly, that egregious animal cruelty is illegal in the United States. They don't know that animals used for food are exempt from the Animal Welfare Act, so they don't know the extent of the cruelty they support at the grocery store. People expect the government to police egregious cruelty, not realizing that the government instead subsidizes it extensively. Sadly, in the United States, the laws can be dictated by the desires of those who make large donations to political campaigns. That is bad news for the animals, because two of the strongest lobbies are the farm lobby and the biomedical lobby—the industries in which the most animals suffer for profit.

Karen Dawn

Horsing Around with the Law

The power of political money over public preference has been on full display during the battle to save America's horses from slaughter.

Tens of thousands of American horses have been dying every year in slaughter-houses. They are destined for dog food or for European fine dining. They are thorough-bred racehorses who no longer win, or horses whose working days are over, or discarded family pets, or the wild mustangs who got in the way of the cattle industry.

Paris delicatessen specializing in horse meat

To protect wild mustangs, activists fought hard, and prevailed, back in 1971. They won a law that banned the inhumane treatment of wild horses and put safeguards in place so that they couldn't be sold for slaughter. As noted in chapter 5 (under "Kill a Horse to Eat a Cow"), however, the cattle industry, which grazes its cattle on the land where the wild mustangs roam, has chipped away at the protective law ever since.

In late 2004, Congress passed an amendment to that law, thereby removing all protection from slaughter for wild horses who are over the age of ten or who have been offered unsuccessfully for adoption three times. Senator Conrad Burns (R-MT) had slipped the amendment into a budget appropriations bill on the eve of the bill's congressional deadline—thus avoiding debate.

Representative Ed Whitfield (R-KY) commented:

> The thing that is so damaging about this Conrad Burns amendment is that he passed it on an appropriations bill that no one knew about. . . . It is precisely the way the legislative process should not work. I don't know his motivations but more than likely he was protecting the ranchers who have leased those lands (for cattle and sheep grazing).[2]

Indeed, an article in *Vanity Fair* tells us that Senator Burns "has a history of being sensitive to the needs of those who donate large amounts of money to his campaign" and "from 2001 to 2006, the senator received $380,512 from agribusiness, which includes the livestock industry."[3]

Since Congress had not set out to reinstitute wild mustang slaughter, legislation to reinstate the federal protection for mustangs quickly passed as an amendment to the House Interior Appropriations bill. Before a bill is signed by the president, however, it goes into a House–Senate conference committee, the purpose of which is supposedly to iron out any differences between the House and Senate versions that arise due to amendments made as a bill progresses. Sometimes conference committee members use the committee for their own purposes. The mustang protection amendment was stripped from the final bill.

In early 2007, Representative Whit-field (noted above for his criticism of the Burns amendment) and Representative Nick Rahall (D-WV) introduced, again, legislation to restore protection to mustangs.

Americans, even most politicians, love horses, so an effort was made to protect all horses, not just wild mustangs, from slaughterhouses. Congress often exerts power by controlling funding under appropriations bills, which must pass, rather than by pushing through new laws. For example, the Vietnam War ended when Congress voted to cut off money for the war effort. Accordingly, during the same period that it was debating the mustang is-

*I can eat one out of a tin but
I couldn't kill one myself.*

sue, Congress passed, overwhelmingly, a bill that denied government funding for inspections at horse slaughter facilities. The measure was meant, as stated by cosponsor Senator Robert Byrd (D-WV), to "stop the slaughter of horses for human consumption,"[4] as the practice cannot take place legally without at least some show of inspections. The Department of Agriculture, however, which we have noted has close ties to the cattle industry (in chapter 5 under "Dangerous Liaisons"), engaged in some creative legal analysis and decided that it could still allow the horse slaughterhouses to operate under federal inspection as long as the funding for those inspections was private. So the slaughter continued until the matter went to the U.S. District Court, which "threw out USDA's attempt to get around Congress."[5]

Only Texas and Illinois had horse slaughterhouses; they are currently not functioning. Illinois has now banned horse slaughter and a Texas court of appeals has voted to uphold a 1949 Texas law banning the slaughter of horses for human consumption. But legislators have introduced a bill to legalize the export of horse meat from Texas for consumption in such countries as France and Belgium.[6] Moreover, while the transport of horses to slaughter is still legal U.S. horses are being transported thousands of miles to slaughter in Mexico.

The Horse Slaughter Prevention Act has been introduced in the hope of finally putting a permanent end to horse slaughter in the United States. It would also ban the transport of horses out of the United States for slaughter. While some legislators fought to prevent the bill from getting out of committee and onto the House floor for a vote, it did eventually make it onto the House floor, and it passed by a landslide. But the Senate failed to take any action on the legislation before the congressional session concluded in December 2006. A senator can hold up the passage of a bill by putting a hold on it. Senator Burns did just that. Perhaps Montana's Burns is starting to remind you of his *Simpsons* namesake, Montgomery Burns.

A *Washington Times* editorial about the debacle referred to "a long list of anti-democratic abuses employed by opponents of the ban to keep them in good standing with their rancher buddies. . . ." It commented, "What we've seen from the cattle ranchers and their legislators is nothing short of a perversion of democracy. Whether one particularly cares about the slaughter of horses, every American should care deeply when lawmakers and agencies obstruct the lawmaking process or choose to ignore the law altogether." Another comment in the same piece was, "If horse slaughter is ever banned in the United States, it won't be for lack of obstruction on the part of supporters of this barbaric practice."[7]

As I write this chapter, the proposed federal ban is being debated. As you read this book, I hope we will be celebrating the ban of horse slaughter in the United States. The cattle-ranching lobby is strong, but the American affection for horses appears to be strong enough to eventually override it.

What happens, however, when we push for legislation to protect animals about whom the public has less passion? Even when public opinion and congressional opinion are on the side of the animals, it can be hard to garner the strength to win at the games politicians play as they protect the interests of those who pay for their political campaigns.

Legislators Bringing Home the Bacon

In 2003, a bill was introduced in California to prohibit keeping calves or pregnant pigs in crates so small that the animals are unable to turn around. A Zogby poll showed that more than 70 percent of Californians supported the legislation.[8] Having passed the Public Safety Committee of the Assembly, the bill should have gone to the full Assembly for a vote, but the Agriculture Committee asked to review it first. They killed it. Though Californians are in favor of the bill, unfortunately their passion for pigs does not have the horsepower to defeat the farm lobby, so the bill went nowhere.

Power to the Poopers

The farm lobby flexed its federal muscles to odious and odorous effect back in 2005, when thousands of livestock companies were exempted from Clean Air Act Requirements in return for paying a minimal fee and agreeing to participate in an air-quality data-collection program. Groups such

Courtesy of Universal Studios Licensing LLLP

After a few months hanging out with Babe, actor **James Cromwell** went vegan. He also became an outspoken opponent of factory farming.

as the Sierra Club and Center on Race, Poverty, and the Environment sued the EPA, but in 2007 a federal appeals court ruled that the EPA was exercising a valid use of the agency's enforcement discretion by entering into agreements with the farms.[9] *Grist* magazine points out that "Of the 14,000 farms signed on to the compliance agreement, only 24 will actually be studied. The rest are happy as pigs in . . . well, you know."[10] I trust *Grist* means nice clean straw! For the record, I have visited many a farm sanctuary and have yet to see a pig rolling around in shit. They do coat themselves smartly in fresh mud to ward off sunburn, but the happy-in-shit myth must be kept alive by factory farmers making excuses for making pigs live in it, and their neighbors breathe it.

Ballot Boxes

Via ballot initiatives, activists sidestep the lobby-driven legislators and take the matter of animal cruelty directly to the compassionate public. Twenty-five states allow ballot initiatives. In various states animal advocates have prevailed, using ballot initiatives against particularly cruel forms of hunting and trapping. Such bills cannot make it through state legislatures due to the lawmakers' unwillingness to alienate the hunting lobby.

In 2007 Oregon became the first state in the United States whose legislature outlawed the confinement of sows in gestation crates. We hope it is the beginning of a trend, but so far other states have had to rely on ballot initiatives to pass similar bans. They have succeeded in Florida and Arizona. After the Arizona win, Smithfield, seeing their shit start to hit the fan, and no doubt trying to slow the momentum, announced that it would voluntarily phase out the crates—over the next ten to twenty years.[11] But that's way too long for the animals to be stuck in it, so a similar California initiative is planned for 2008. Get out and vote!

Living with Electile Dysfunction

Unfortunately, we can't rely completely on ballot initiatives—not all states have them, and there are no federal ballot initiatives. Another worthwhile tactic for animal advocates is to donate to political campaigns of animal-friendly candidates. It is hard, however, to compete with the wealth and strength of the biomedical and farm lobbies. For that reason campaign finance reform is a critical campaign issue, arguably the most critical, for the animals. Until there is significant reform in that area, however, those who care about animals must find nonmonetary means by which to influence politics.

We are moving in the right direction. On Election Day 2006, the *Wall Street Journal* put "Puppy Power" on its front page.

The article told us:

> For the first time in its 50-year history, the Humane Society is trying to elect candidates to Congress who support its animal welfare agenda. After a series of mergers with other animal welfare groups, the Humane Society counts 10 million Americans as members, an average of 23,000 in each of the 435 House districts. That's more than twice the membership of the National Rifle Association, which is considered one of the most effective single-issue campaign organizations.[12]

The comparison to the NRA is ironic but important. Its members constitute a small proportion of the population who wield enormous power. When I think of the NRA I think of the spiritual lecturer Marianne Williamson's comment about worldly power. She says the problem isn't that there are so many people in America who align themselves with darkness, "the problem is that they get up so early in the morning."[13] They are organized.

The NRA's members, who account for just a few percent of voters in most districts, vote as a block and can swing elections. Those who love to hunt vote for a candidate not according to whether that candidate is a Republican or a Democrat, but according to the candidate's position on guns. Only if both candidates are pro-gun will NRA members consider other issues.

We can't yet imagine that the 72 percent of Americans who say they support the goals of the animal rights movement will take that support to the point of voting according to a candidate's record on animal issues. But the Humane Society of the United States (HSUS) is trying to persuade its members, averaging tens of thousands per district, to do just that. If those of us who care deeply about animals make it clear that we will vote accordingly, we will have the power to swing elections.

See Ya Later Animal Hater

Senator Conrad Burns, infamous for the last-minute rider that reinstated wild mustang slaughter, was in one of the tightest races of the 2006 election. HSUS contends that its flurry of ads, which reminded Montana voters that Burns was almost personally responsible for the slaughter of wild mustangs, may have cost Burns the election. A few thousand animal lovers at the polls are enough to make the difference.

Crossing the Line to Save Lives

Animal protection has traditionally been seen as a Democratic cause. That reputation is unwarranted. It is true that Democrats tend to have better voting records on animal issues, but the exceptions are glaring. Former Senator Robert Smith (R-NH), an ultraconservative Republican, is the only person to date to speak passionately against vivisection on the Senate floor. Republican John Ensign of Nevada has been a leading voice in the battle against horse slaughter. Representative Chris Shays (R-CT) delighted us with his rousing speech at the 2007 Taking Action for Animals Conference.

A beautiful book on animal protection, *Dominion: The Power of Man, the Suffering of Animals and the Call to Mercy*,[14] was penned by Matthew Scully as he served as a senior speechwriter to President George W. Bush. Scully reworked a section of that book into an article headed "Torture on the Farm," and Pat Buchanan put it on the cover of his magazine, *The American Conservative*. We are still waiting for articles on farm cruelty to grace the covers of such bastions of liberalism as *Mother Jones* and *The Nation*.

Some members of the animal rights community have been distressed to see animal-protection groups back candidates whose stances are animal-friendly but unfriendly to other social justice causes. Obviously, many of us care deeply about issues other than animal rights. Moreover, some activists think that in order for our movement to succeed, we must align ourselves with other social justice movements. But while those movements are still serving factory-farmed meat at their functions, is it appropriate for us to vote for their candidates if one of their opponents takes a stand for the animals? That will sometimes be a hard question for me, as I do not care more about animals than people. But the animals cannot vote for themselves and have no hope of advancement until a group of people is willing to say that they matter enough for us to vote according to their needs. Having given them my word that I will help them, I cannot vote against them.

The good news is that when we start to show up consistently as a voting block that can swing elections, we will not have to vote against our other ideals. Candidates on both sides of the aisle will have animal-friendly policies. We will have modeled not our ideals but our behavior on the NRA, which has made sure that candidates in both parties currently support the right to hunt. For now, we understand Bill Maher's exasperation as he discusses the candidates' hunting photo opportunities and he asks, "What is it about this country, that a man can't run for president if he is insufficiently cruel to animals?"[15] Won't it be nice to see, as standard political campaign footage, candidates from both parties walking into the woods with rehabilitated animals they have rescued, rather than walking out of the woods with animals they have shot?

Letters to the Legislature

Whether your government representative is known to be animal-friendly or not, your voice can make a big difference. When bills come up for a vote, legislators take seriously the feedback they have received from the public. They pay particular attention to notes and calls from their own constituents, who they have, after all, been elected to represent. (Though few legislators would disdain attention from all over the country—particularly if they have their eye on a big federal prize.) A few dozen notes or calls—for or against a bill—can make the difference for the vote of a legislator who is on the fence. Because sentiment does not equal political power, animal advocates should be aware of animal issues that are before the legislature and should get involved. The Humane Society Legislative Fund's Web site, www.Fund.org, is a great resource. Or sign up for an alert list at community.hsus.org/humane/join. Remember that without activism in the political process, the balance will always tip in favor of the organized business lobbies.

If You Can't Beat 'Em, Join 'Em

Generally animal advocates wish to reach out to compassionate people involved in other causes. Those people are already demonstrating an interest in a kinder world.

Things can get sticky when people are supporting charities, in the hope of helping people, without knowing how much those charities are hurting animals. They may not know, for example, that many medical research charities devote their money to outdated and ineffective animal research that is under-monitored for issues of egregious cruelty. Therefore, when we protest against the March of Dimes, for example, the general public might see our protests as misanthropic. We will be better understood if we focus, in such a protest, on the now infamous March of Dimes study in which the eyelids of kittens were sewn shut and left that way for months. Or we might focus on March of Dimes funding for studies in which nicotine, cocaine, and alcohol were given to pregnant animals, even though we already knew from human clinical experience that those substances could harm a developing baby. Unless we make sure that the public learns what we know about the organizations we protest against, we run the risk of alienating compassionate people who are the most likely to become great friends to the animals.

We can learn from the mistakes of others. In Scotland, the leading charity for children with learning disabilities designed posters telling people that animal charities attract almost twice as many donors as disability charities.[16] The campaigners set up conflict between causes, and alienated people very likely to be sympathetic to children. A leading animal advocate, who had been put on the defensive, responded to the campaign with the statement, "They forget that the welfare state was established to ensure that the educational and health needs of the people of this country would be paid for through our taxes. There is no welfare state for animals. There is no 'Children in Need Day' for animals. People who choose to support animal charities have already paid for the welfare state through their taxes."[17] Of course, children with learning disabilities could use even more money than the government is sending their way. Attacking animal advocates, those who have demonstrated a willingness to give to those whom society has forgotten, is probably not the wisest way to try to attract it. And animal advocates should think carefully and strategically before going after those who support charities that hurt animals.

On Hating Heifer

We find ourselves, as animal rights advocates, in that uncomfortable situation when we see groups such as the Heifer Foundation encourage gifts of live animals to people in developing nations. In *The Independent* (London), Andrew Tyler, the director of Animal Aid, presented some compelling arguments against such gifts. He wrote:

> It is incontestable that desertification and further human impoverishment will follow the introduction of goats into already degraded areas. But if goats are environmentally disastrous, cows are extraordinarily burdensome economically. A newly lactating animal requires up to 90 liters of water a day, a lot of food and veterinary treatment to cover endemic problems such as scours, mastitis and lameness.[18]

It is hard to imagine people in the developing nations being able to provide veterinary care. Even providing adequate shelter will be a burden. And the issue of water is crucial.

Tyler wrote:

> It is many times more efficient to use the available agricultural resources—land, labour, water—to feed people directly, rather than devoting those resources to fattening animals.

There are, however, issues being left out of the discussion. In chapter 5 (under "Taking Care with Vegan Fare"), I discussed the many advantages of plant-based diets, but I also noted that those benefits are for those of us living in developed nations with easy access to nutritious food. Westerners tend to be overnourished and overweight rather than undernourished or starving. The Kenya study I cited in that section showed that a small amount of meat, when added to inadequate diets, brought health improvements. While adding cow milk had no better immediate impact than the same calories in vegetable oil, it did provide vitamin B_{12}, a necessary nutrient for good long-term health, and one lacking in the plant-based diets of those in developing nations.

I raised the B_{12} issue with Andrew Tyler of Animal Aid. In his response, he wrote about vegans: "As a group we certainly compare favorably with meat and dairy consumers—many of whom consume a thoroughly unhealthy and nutritionally deficient diet and exhibit all kinds of neurological and other disorders."[19]

True—but not for the kids in Kenya.

With regard to B_{12} he suggested: "Many of the foods we eat are fortified with B_{12}; the aid charities could resolve to work with the appropriate agencies/local businesses to fortify a food that is a staple in the given area."

That is a superb suggestion. Animal groups should, indeed, encourage the fortification of local staples with B_{12} and omega-3s and other important nutrients.

Tyler's basic premise is spot on. Encouraging animal agriculture will not, ultimately, help the developing nations. Let's not forget Edward O. Wilson's well-researched contention (covered in chapter 7 under "Little Piggies") that the amount of grain the world produces could support 10 billion vegetarians but only 2.5 billion people who convert that grain into animal protein.

In campaigning for the cessation of animal donations, however, without appropriate alternatives in place, we may be premature. While working to make sure fortified, nourishing vegetarian options are available, it might be best not to protest donations of those animals who can be best cared for, who will do the least environmental damage, and who provide the greatest nutrition over a long period of time. Perhaps currently, rather than protesting their donation, we should insist on welfare guidelines for them even as we point out the environmental consequences of their introduction and we prioritize our work with relief agencies to supplement staples so that animal products become unnecessary. We owe it to

This is all about "thanks." Next month, the massacre starts all over again in the name of "peace on Earth."

the animals to demonstrate that caring for their needs does not mean caring less about people; otherwise, our efforts on their behalf will be ignored. And we must reach out to and work with, rather than against, all others who are attempting to put compassion into action.

media matters

Befriending the media can help us reach out. The power of the media is awesome. The media tell society what is hot, what is hip, what matters. The media can be used for great good. Films like *Philadelphia* and shows like *Ellen* and *Will and Grace* have surely had a profound positive effect on mainstream America's view of homosexuality. Those of us who advocate for animals can probably thank *Flipper* for making the very thought of eating dolphin meat nauseating for Americans. News media, too, can expose hidden cruelties and change people's habits. Those who saw the *Dateline* puppy mill piece, for example (discussed in chapter 2, under "Paritney's Pet Store Adventures"), would surely have been reluctant to get their next animal companion from a pet store.

It is up to us, who wish to help the animals, to befriend the media on their behalf. While the media are incomparably influential, they are surprisingly easy to influence. Journalists, editors, writers, and directors, who control what is in the media, are just people who do stories on what they find interesting and to which they think others will also respond. They cannot cover anything they have never heard about. If a reporter knows nothing about the cruelty of the dairy industry, she isn't going to cover it.

Those who work in the media also want to know that their stories are going to be popular, as popularity translates into high circulation or ratings and into advertising dollars. That's why feedback to the media can have so much power. How much power? When ABC *World News Tonight* producer Judy Muller accepted a Genesis Award for a series of animal-friendly reports, she told the audience, "The good news is that the audience response to these stories is wonderful, and that means that these stories will keep coming. In the hard-hearted world of network news, that's the bottom line."[20]

In chapter 6 of this book I referred to a superb episode of *Boston Legal* in which the Draize Eye Irritancy Test was described in detail for viewers. When the show aired, I sent

INVESTIGATE VEGETARIANISM

—JORJA FOX

PeTA FOR A FREE VEGETARIAN STARTER KIT, CALL 1-888-VEG-FOOD OR VISIT GOVEG.COM.

Poster compliments of Jorja Fox and PETA.

CSI actress **Jorja Fox** knows the power of the media. Her character, Sarah, was a veggie gal—just like Jorja.

information about it to my DawnWatch list subscribers and asked them to please post positive messages on the *Boston Legal* message board and to send appreciative e-mails via the ABC feedback page. It is important to let stations know that animal-friendly material has an eager audience. I was gratified to receive a note from director Eric Stoltz. He let me know that the people on the set had been largely pro–animal rights, but he also wrote, "Thanks for pointing out the need to contact ABC. They are the final arbiters of taste and editing and hold all the cards re content."

Animal-friendly people in the entertainment industry get license to produce work close to their hearts when they can demonstrate audience interest in their issues.

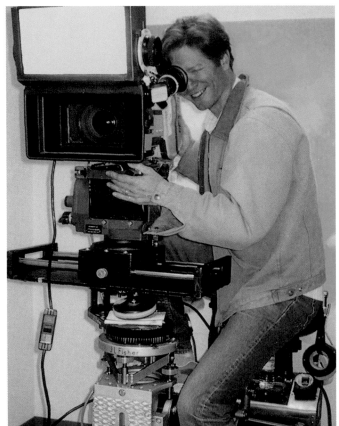

Photograph compliments of Eric Stoltz

Eric Stoltz directing

I launched DawnWatch.com to help keep animal rights activists abreast of animal issues in the media and to encourage them to respond. Having worked as a researcher and writer on Australia's national news show *The 7:30 Report*, I already had a basic understanding of how information gets to reporters and how news shows are shaped. Then I was inspired by a personal experience in which I saw the power of feedback. People who have heard me lecture, or read my chapter in *In Defense of Animals: The Second Wave*,[21] will be familiar with the story. I will share it here in the hope that it will inspire others to get and stay in touch with their local media.

Media Circus

When I lived in New York I regularly watched the twenty-four-hour news station NY1. Over the weekends NY1 plays about three hours of programming in a continuous loop, the same shows airing over and over. One of the shows on the loop was called *Parenting*. Every week anchor Pat Kiernan, who you may now also know as the delightful host of VH1's *World Series of Pop Culture*, interviews Shelley Goldberg from *Parenting* magazine. Shelley gives recommendations for fun things to do with the kids on the weekend.

One Saturday morning, Shelley said that Ringling Bros. was in town; she told New Yorkers they should take their children to the circus to see the lions and the tigers and the elephants. I had been involved in animal rights for just a few months, but I had already seen the footage (described in chapter 3 of this book under "An Elephant Never Forgets a Bullhook") of baby elephants being beaten with bullhooks by circus trainers. I realized, with dismay, that many thousands of New Yorkers had heard Shelley's recommendation, and that it would replay many times over the weekend for thousands more to hear.

I then remembered that a friend who hosted a show on NY1 had told me that the station manager reads every single e-mail that comes in. I immediately sent an e-mail to the station. I praised the station's overall work but expressed surprise that it had covered the circus as a fluff piece rather than as a controversial news story.

Then I sent a note to the activist Susan Roghair, who I knew had an extensive e-mail alert list. I asked her if she could alert her New York subscribers to the recurring circus story. Within minutes I received, as one of her New York subscribers, an e-mail telling me about the circus story and asking me to contact the station. New York has some terrific activists campaigning against the circus and clearly quite a few of them got in touch with the station. I had NY1 on in my apartment for the rest of the weekend, and for the first time in my memory of that station, the *Parenting* segment was taken out of the weekend loop.

Nice! Then things got even better.

On weekday mornings, the same anchor, Pat Kiernan, did a regular segment called *In the Papers*. He would look through the New York dailies, read the headlines, and then choose one or two stories from within the papers on which to focus. The Wednesday following the *Parenting* circus, Kiernan chose a story from deep within one of the tabloids: He told his viewers that Ringling Bros. had been cited for noncompliance with the Animal Welfare Act.

How gratifying to see the anchor who had received the feedback on circus promotions make sure, the same week, that NY1 morning viewers were aware of a somewhat buried story on circus cruelty.

On the following Saturday's *Parenting* segment, Shelley Goldberg recommended we take our kids to the puppet show. Then she looked at the camera and said, "We want all of our viewers to know that no animals were harmed during the making of this puppet show."

Not only did feedback from viewers change the programming on the weekend when the circus piece had been scheduled to air continuously, it also affected two media personalities. It made them sensitive to the controversy surrounding the circus and aware that their viewers care about the issue.

That example focuses on constructive criticism. Any criticism must be delivered in a way that is unlikely to alienate the reporters. The media are powerful and the animals need powerful friends. We can also, as the ABC producer Judy Muller noted, have just as strong an impact with completely positive feedback. This book's next section provides an example.

RUBES®   **By Leigh Rubin**

**Early radicals of the Animal
Rights Movement**

Karen Dawn

Slaughterhouse Five TV

There is a saying in television news, "If it bleeds, it leads."

If only that were true for the animals. Sadly, most news stations refuse to air any graphic footage of animal suffering. They say people find it too disturbing and will just change the channel and watch the station's competitors.

A brave reporter named Duane Pohlman, however, managed to talk the NBC affiliate King 5 in Seattle into airing some of the horrifying undercover slaughterhouse video that I discussed in this book's chapter 5 section "Dying Piece by Piece." He told viewers,

> By law, cows are supposed to be unconscious when they're slaughtered. But workers say some of these animals are very much aware of what's happening as they're butchered. . . . Every day, animals pass down the belt alive.[22]

The segment included interviews with workers who described cows blinking, mooing, and kicking as they are cut up. Viewers saw the footage.

The story was available online on the station's Web site, so those in Seattle who missed it, as well as activists all over the country, could see it. Media-sensitive activists urged people to thank the station for airing the story. When a follow-up was aired, the anchors noted that the station had received hundreds of e-mails in response to the first story. The follow-up received more positive feedback. The station aired over a dozen stories on the IBP slaughterhouse violations, all showing some of that graphic footage.

Paul McCartney is reported to have said, "If slaughterhouses had glass walls, everyone would be vegetarian."[23] I suspect a lot of people became vegetarian in Seattle that month.

Later I spoke with Pohlman, now an Emmy-winning reporter, about the effect of the audience feedback on the coverage. He told me that the station had been reluctant to air even the first segment, and said, "Without the overwhelming e-mail response it would have been highly unlikely that the station would have committed to the number of follow-ups it did. It gave the station footing, and me coverage, to continue. I could say, 'Look at this—people are moved about this, even people across the country. It is incumbent on us to follow up.'"[24]

"If slaughterhouses had glass walls, everyone would be vegetarian" (**Paul McCartney**).

Though it is attributed to him in dozens of places, Sir Paul has no actual recall of having uttered that line. He thinks it might just as easily been have Linda, or even somebody else. But he is a well-known animal advocate and is happy to have his photograph included in this section of our pro-veggie book.

Sebastian Smee, "Beastly Goings On," *Weekend Australian*, March 10, 2007, p. 18.

The Power of One

While the stories above demonstrate the results of a concerted effort, a single activist can have a powerful effect simply by developing a good relationship with a reporter. The obvious reporter to target is one who has done stories or made comments that indicate some sympathy to animal issues. We can surprise ourselves, however, with the inroads we can make elsewhere. A terrific activist who focuses on rodeos, Lucy Shelton, shared the following with me: When Lucy saw an article in the *Inland Valley Daily Bulletin* promoting the rodeo, she called up the reporter, was pleasant and friendly, but told her what happens to the animals on the rodeo circuit. She asked if she could e-mail the reporter some more information. The very next day the same reporter did a story headed "Animal Rights Group Heads for Rodeo." That article mentioned that some of the rodeo cowboys had been cited for animal abuse, including the use of electric prods on horses. With just one phone call to the media, one activist made sure the public learned about the dark side of the rodeo being held in her town.

Animal Love Letters

Those who care about animal issues should not underestimate the power of letters to the editor. A study of 296 newspapers reported, "Large- and medium-sized papers that have conducted surveys consider the letters to the editor column to be 'one of the best-read items.'"[25] No newspaper reported the letters column was below average in readership. Moreover, legislators look to the letters pages as barometers of public opinion. Your letter can let legislators know that animals matter to their constituency.

Even if your letter does not get published, it will be read by at least one powerful person—the letters editor. The more that person learns about animal issues the better. Also, if a paper receives many letters in response to the same story, generally at least one will be published. It is important that those many letters do not have similar wording, which is why sample letters to the media are not useful. Editors want to hear from their readers, not to feel that they are the victims of campaigns. When you send an original letter, however, whether your letter or another letter on the same topic is published, your effort will have helped to give the animals a voice.

DawnWatch

I always encourage animal advocates to sign up for DawnWatch, my free media alert service. DawnWatch alerts encourage letters to the editor and feedback to stations that cover animal issues, because positive feedback for animal-friendly coverage encourages more of it. DawnWatch focuses mostly on national media, but it is a good training ground where activists can learn how to respond to all media so that they can take those skills to their local media outlets. At DawnWatch.com or ThankingtheMonkey.com, you can sign up for DawnWatch alerts.

"I am writing with regard to alcohol and nicotine tests on animals."

© Jeffrey Fischer

From Armchairs to Sidewalks

The animal advocacy described above, where one sits at one's computer or on the phone and contacts the media, or contacts legislators, is sometimes called armchair activism. Though the name has slightly derogatory connotations, it is effective. So are other forms of activism.

Some people go on demonstrations. They protest outside stores that sell fur, or outside laboratories where animals are used in experiments, or at zoos and circuses. Because media coverage can be what makes demonstrations effective, activists must attempt to create events that are newsworthy and camera-worthy. One activist in a rat suit is worth twenty in jeans, because interesting visuals attract the press. Actually, most outfits are better than jeans—the better groomed you are, the better for the animals.

I remember arriving one morning at a national animal rights conference that was held in Los Angeles and being told there was a huge picture of me in the *Los Angeles Times*. Actually, it was a huge picture of Paula, with the caption "Paula Dawn, a pit bull terrier, gets comfortable while her human companion Karen Dawn speaks during the animal rights conference."[26] The picture also included a very small image, thankfully, of my backside, because Paula's close-up was taken as she lounged behind me on the stage. There were dozens of speakers at that conference but I was the only one there with my dogs—and they provided the good photo opportunity. Because of the size of the photo, the story about the animal rights conference took up almost half a page in the *Los Angeles Times*.

Another time our family was interviewed for a story specifically about pit bulls. When the photographer showed up, I threw Paula's "doggles" on her, which look both ridiculous and cool. The story made the front page, no doubt because Paula looked so fabulous that she also made "The Week in Pictures." Yes, you are reading the gloatings of a proud stage mother, but there is an important point here: An unusual visual got a pit-bull-friendly story onto the front page of the *Los Angeles Times*.[27]

Though it can get annoying when we have greater issues on our minds, activists have to think about the good shot, and about the story the media will want to cover. Television, in particular, is all about visuals. If a demonstration makes the media, it won't have mattered much if there were five or twenty activists at the protest. What will have mattered is that they got media to their protest and therefore got widespread coverage of their issue. Then it will matter that the activists at the protest were articulate—or better yet, that an articulate spokesperson was selected—and that the media were given good information on the issue being protested.

Running of the Nudes

Nobody knows all of that better than PETA. PETA is famous for pulling stunts to get press. Once the cameras and microphones are turned their way, they talk seriously about the issues. A perfect example is the PETA Running of the Nudes, a yearly alternative to Pamplona's Running of the Bulls. PETA's human and humane version of the race is racier than the original! When Bill O'Reilly invited PETA vice president Lisa Lange onto his show to discuss the event, he asked, however, if it was basically a protest of the bull killing. Lange said, "Anyone who watches the race and sees these animals as they're slipping and sliding and being forced to run down the streets has to see how terrifying

that is. But what most people don't know is that every single one of those terrorized bulls is killed in the bullring that night and through the week." She reminded us that the majority of Europeans, even Spaniards, are opposed to bullfighting. (Indeed, *Time* magazine tells us that, according to a Gallup poll, only 8 percent of Spaniards consider themselves bullfighting fans.[28] Barcelona even declared itself a non-bullfighting city in 2004, though unfortunately the vote was nonbinding as the regional government of Catalonia has jurisdiction.[29]) Lange discussed bullfighting in some detail: "What happens in every bullfight is that a dagger is stuck in the muscled part

Andy Sims/AndySimsPhotography.com

of the back of the bull's neck, and they're eventually stabbed in the back. Many of these animals bleed to death. They're paralyzed but they're still fully conscious and able to feel pain when they are dragged out of the ring. It's a very violent blood sport and it's time to see an end to it."

Then O'Reilly's fans watching the show got to hear O'Reilly's take on the issue:

I think it is like dog fighting and other sports where you have an animal that's going to get killed for the entertainment of people— there's something disturbing about it.

Viewers never would have learned O'Reilly's opinion of animal blood sports if it had not been for the PETA Running of the Nudes segment.

O'Reilly then told us all something PETA seems to have worked out some time ago:

You guys always use the same technique, let's get naked, then we get media attention. And it works. I mean we like you as a guest and all that, but I don't know if we'd be doing a segment if there weren't lots of naked people in Spain. We might, but it was assured once we got the naked video that we'd do the segment. I want to be honest with the audience, that PETA does this to get attention and it works.

You might look at the folks pictured on these pages and think PETA would have better luck if they stopped sending in the vegans and threatened instead to have McDonald's executives walk naked through Pamplona unless the bullfighting stops. But that wouldn't get on O'Reilly and give two million viewers information likely to make them rethink their amusement choices on their next vacation.

Andy Sims/AndySimsPhotography.com

freedom fighters

Animal Liberation Front (ALF) activism seems more glamorous than letter writing and demonstrating—though there is nothing glamorous about jail. The ALF are the freedom fighters for the animals. They risk their own freedom to go into abusive situations—commonly laboratories or fur farms—and rescue animals. They get immediate gratification; they break locks and see once-caged animals run free or rest in loving arms in loving homes.

Obviously the ALF helps those animals they free. But what of the animals who are bred into a life of suffering to replace those taken from the cages? Does the suffering of the replacements matter less than that of those who were freed? Because animal cages will immediately be refilled with new animals, one has to question whether actions that free individual animals help animals overall. Some in the ALF would argue that their actions put a dent in the profits of animal-cruelty enterprises. Often ALF activists, besides removing valuable animals, will also intentionally damage property at institutions they visit in order to increase that dent. But given the enormous profits of animal enterprise, will trashing one lab, one store, or one farm make that much difference?

From the Front Line to the Front Page

It depends on the press. In 2003, militant activists trashed a restaurant in Northern California that was soon to open and specialize in foie gras. The $50,000 worth of damage was surely a blow to the store owner, but thanks to insurance, probably not a deal breaker. The activists got lucky, however, when a local reporter, ABC's Dan Noyes, picked up the story and helped change the course of history for the animal rights movement. He decided to look behind the actions of the activists to the suffering of animals on foie gras farms. He aired a story showing gruesome footage of foie gras production.[30]

The public responded favorably. Noyes tells me the station received hundreds of e-mails and that "the contact from the public reinforces that we are reporting stories our audience finds important."[31] So the station aired six stories, all including graphic foie gras footage. That coverage, throughout the Bay Area, set the stage for a bill, brought by San Francisco's Senator Burton, to ban the production and sale of foie gras in California.

The bill passed and will take effect in 2012.

It was not the economic damage inflicted by the act, but rather the damage in public opinion inflicted by the media, that made the difference. Some activists, acknowledging the power of the press, have adjusted their actions accordingly. An Australian activist named Patty Mark is widely acknowledged for having brought "open rescue" to the world, a tactic that such groups as Compassion Over Killing and Compassionate Consumers have popularized in the United States.[32] Open rescuers do what the ALF do— they go into animal enterprises, such

"I'm sorry, sir. PETA just rescued your dinner."

as factory farms, and they rescue animals. But while ALF activists wear masks to hide their faces from security cameras, open rescuers film themselves at work, mask-free, then deliver the film to media in order to shine the spotlight on institutionalized cruelty. A perfect example of an effective open rescue was that carried out by Adam Durand and other members of Compassionate Consumers at the Wegmans Egg Farm. Adam spent thirty days in jail but the video he took was covered by the *New York Times*[33] and aired in a segment on ABC's *Primetime*,[34] teaching millions of Americans about the suffering of egg-laying hens. It is reasonable to assume that some of those viewers switched to cage-free eggs. Others might have suddenly better understood their vegan friends, and lost their taste for eggs. That is effective activism!

Some mainstream animal advocates contend that militant activism is not only of questionable value but that it actually hurts the animals by hurting the reputation of our movement. In 2004, I contributed a chapter to the anthology *Terrorists or Freedom Fighters: Reflections on the Liberation of Animals.*[35] My chapter, titled "From the Front Line to the Front Page," focused specifically on the media coverage received by militant action. I made the following points: The media have largely ignored animal issues, and silence serves the status quo. At least the militant activists get press. With letters to the editor, activists can help change the slant of that press, taking the focus off the activists and putting it onto the animals, as we have done effectively in the past. When militants get press, those of us who have vowed to help the animals cannot ignore it. We must do what we can to shape that press. We should present our movement as peaceful but devote more energy to disavowing the animal abusers than to disavowing the activists.

I did my best, in that chapter, to make my take on the militant activism itself even-handed. Every social movement has had militant and unpopular activists, and we do not know if those movements succeeded in spite of or because of them. A good cop–bad cop phenomenon surely comes into play, with the public and legislators feeling that if they don't take the mainstream activists seriously, they are going to have to deal with the militants.

Outsourcing Tech Support and Torture Support

A phenomenon that has come to attention in the years since I wrote that chapter is the outsourcing of animal testing to China, where, according to the *Boston Globe*, "scientists are plentiful but activists aren't."[36] Neither are animal welfare laws, and that concerns animal advocates, some of whom suggest the militant activists are making it worse for the

animals by driving testing to China. But if you are being poisoned to death with oven cleaner, does it really matter how big your cage is and in what country?

Something to consider is that animal testing persists because it is a multimillion-dollar industry. There is much money in the breeding

© Andrew Weldon

and sale of animals. If it were a multimillion-dollar Chinese industry, then U.S. lobby groups that make millions from animal testing wouldn't interfere with U.S. activists' efforts to change U.S. testing requirements. We should therefore question whether the outsourcing of animal testing to China is a negative. By removing some of the financial incentive of the U.S. biomedical industry to lobby for animal testing, it might open the door for the approval of testing methods that do not use animals.

Militant Mishaps

In just the last few years the media has become willing to cover the plight of the animals without violent activism's attention-grabbing stunts. Various other events, outlined below, have also nudged me off the fence toward a stance for peaceful activism.

Doug Hall/PCRM

Grave Consequences

For years UK animal rights activists campaigned to close down a Newchurch guinea pig farm. The animals there were bred and sold for testing, mostly of household products. According to the *Sunday Times*, the relentless campaign included hate mail, malicious telephone calls, hoax bombs, and arson attacks.[37] The grand finale was the disinterment of the body of a grandmother who had died at age eighty-two, who was part of the farm-owning family. With that, the family gave up and said they would farm guinea pigs no more.

A win for guinea pigs? Not if another farm provides the animals for the same tests. Perhaps a win, however, for the animal abusers, as they gained much ground in the battle for public opinion. The grave robbing mortified the public. Our movement forfeited the high ground. The *Daily Telegraph* in the UK conducted a survey and published its findings under the headline "Public Turns on Animal Terrorists: 7 in 10 Back Live Medical Experiments."[38] The *Guardian* Web site covered the survey with a Press Association story headed "Animal Activist Campaign Backfires."[39] That article noted that the 70 percent support shown for animal testing was up from earlier polls that showed a fifty-fifty split. Many articles suggested that militant activists had caused the public to lose sympathy for the movement; the articles focused particularly on the grave robbing.

Does the loss of public sympathy matter as much as the closing of the farm? Absolutely. Whereas the experimental animals will simply be bought elsewhere, hard-won public sympathy is not so easily replenished. Public opinion guides public behavior—what people choose to eat and wear—and also guides legislators, particularly in the UK, where the lobby groups have less influence.

SHAC Backlash

In the United States, too, we saw backlashes when activists behaved with seemingly little concern for public opinion.

The Stop Huntingdon Animal Cruelty campaign (SHAC) began in the UK. It ignited when undercover video taken at the Huntingdon Life Sciences (HLS) laboratory showed a scientist punching a beagle puppy in the face as punishment for squirming in fear. The video also showed a monkey on an operating table lifting her head, conscious, with her chest cut wide open. Many people, perhaps having heard about the supposedly stringent welfare laws regulating animal testers, might have assumed such shocking cruelty would have led to the company being permanently shut down by the government. No. HLS got a slap on the wrist from the UK government. That government then propped up the company as activists attempted to put it out of business. The activist effort drove the company out of the United Kingdom to the United States.

That activist effort, the SHAC campaign, has included what is known as tertiary targeting: that's when activists encourage companies to cut financial ties to the targeted company so that it cannot function. Those who refuse are harassed. Harassment has included threats of violence against family members of employees of banks and insurance companies.

Seven activists, later dropped to six, were charged and convicted for running the Web site that was alleged to encourage acts against those who do business with HLS. They were charged under the Animal Enterprise Protection Act, a 1992 piece of legislation that reflects the strength of the animal enterprise lobby. The law singles out crimes—such as property damage—against businesses that use animals, and treats those crimes more seriously than similar offenses against other organizations. During the period of the trial, the Committee on Environment and Public Works held "a second hearing on eco-terrorism specifically examining Stop Huntingdon Animal Cruelty—SHAC."[40] Legislation that would strengthen the Animal Enterprise Protection Act into the Animal Enterprise Terrorism Act had been proposed.

Senator Frank Lautenberg, a New Jersey Democrat, was on that committee. In an article written by a vivisectionist who had his laboratory trashed by animal rights activists, I had previously read these scathing comments about Lautenberg's attitude, which had predisposed me to him favorably:

> Incredulous of the testimony provided by the FBI and the Bureau of Alcohol, Tobacco, Firearms and Explosives (ATF), in which violent animal rights and environmental extremists were identified as among our most serious domestic terrorism threats, Lautenberg asked facetiously who the next target would be: "Right to Life? Sierra Club?" Then, he inexplicably proclaimed himself "a tree hugger."[41]

Lautenberg has supported much animal-friendly legislation. For example, he introduced the Pets Evacuation and Transportation Standards Act, requiring state and local disaster-preparedness plans to take into account the needs of people with pets and service animals.

In his opening remarks at the committee hearing on SHAC, Senator Lautenberg said he supports animal rights, making it clear he is a legislator who at least intends to be a friend to our movement. He said he believes that animal testing is still necessary but he looks forward to the day when it is not. He said he respects the rights of those who disagree with that stance to engage in public protest and legitimate debate. But he said that he deplored the violent, criminal, and threatening activities of the Stop Huntingdon Animal Cruelty campaign since those actions are used to "tar with a broad brush anyone who supports the cause of animal rights or protecting the environment."

"Close down the m*therf@ck$rs!" (**Richard Pryor**, on the Huntingdon Life Sciences animal-testing laboratory).

Available online at www.RichardPryor.com.

Lautenberg, however, reiterated his belief, as portrayed in his quote above, that our nation has much greater threats than animal rights activists. Then, on listening to the testimony of those from the biomedical industry who sought to strengthen the Animal Enterprise Protection Act, Lautenberg suggested that conventional law enforcement codes might already protect people from the actions described. He also expressed concern about the proposed upgrade, asking, "If a boycott was threatened and the company's stock dropped on the Exchange or in the marketplace, would that be included in the recommended statute as loss of economic value?"

The California surgeon and Animal Liberation Press spokesperson Dr. Jerry Vlasak testified. He cited the frequent horrendous abuse of animals at Huntingdon Life Sciences, and by others in the biomedical industry. Vlasak's testimony, however, was tainted by comments he had previously made at an animal rights conference—he had suggested that the murder of a few vivisectors could save millions of animals. When questioned on those statements, he stood by them. Lautenberg did some of the questioning. Listening to the hearing, one can almost hear Lautenberg's blood pressure rise as his conversation with Vlasak progresses.

Lautenberg called "dastardly" the SHAC campaign acts defended by Vlasak, and said he found Vlasak's views on murder "revolting." Vlasak stood firm, presenting his views as he saw right, seeming to have little interest in delivering them in such a way that might lessen his audience's alienation. Finally, Lautenberg, in a rage, said, "I don't want to waste my energy anymore."

Lautenberg became one of the sponsors of the Animal Enterprise Terrorism Act. That act was protested by mainstream groups, such as HSUS, and even by the American Civil Liberties Union. Ironically, the resistance was over the very language about which Lautenberg had expressed concern, which could be interpreted to make illegal any boycotts or protests that are effective. My impression was that Lautenberg's vehement rather than measured support of the act was the direct result of Vlasak's testimony.

Vlasak soon expounded similar ideas on CBS's *60 Minutes*, saying that vivisectors should be stopped "by any means necessary,"[42] even when he was questioned about the possibility of murder. Monitoring animal rights activism discussion lists during that period, I saw much outrage and distress in response to his statements. Could there have been any positive effect? One might argue that people tend to think of animal protection as a morally righteous but somewhat trivial issue, and that Vlasak's stance—a willingness to talk of killing people for the cause—gave the cause a more serious image in the minds of some viewers. Those viewers with whom I spoke, however, thought Vlasak reprehensible. They also seemed to react with visceral distaste for the animal rights movement—a movement for which they had previously felt more tolerance.

There were no polls taken regarding attitudes to animal rights issues among members of the public before and after they saw that *60 Minutes* piece. I can only suspect that if such polls had been taken, they would have shown a shift away from the animals, similar to that seen in the UK in the wake of the robbing of a grandmother's grave.

My point is not about the morality of Vlasak's statements. While many of us might question the general contention that it is acceptable to take human life to save animal life, we nevertheless might pick up the nearest baseball bat and slam it against the skull of anybody torturing to death our beloved dog. I noted in chapter 2 of this book that most of us could think of some person over whose life we would choose our dog's— even though all things being equal we would naturally favor human life. The point, however, is that as activists we must never just shoot off our mouths, letting the animals take the hit. We owe it to them to say things in a way that other people will be willing to hear, and to think about whether what we say will be useful and truly helpful to those we have vowed to help.

© 1989, 1994 by Berkeley Breathed. Distributed by The Washington Post Writers Group. Reprinted with permission.

A SHAC Retraction?

It is interesting to note comments, during the same period, made by Kevin Jonas. Jonas had been the previous president of SHAC USA, and was awaiting sentencing. While he told a *Toronto Star* reporter that he was not legally responsible for the crimes of which he was accused, he also said:

> If I had to do it over again, I'd censor more of those Web postings and be less aggressive and confrontational, so that no one could perceive us as a threat, only as an embarrassment.
>
> We want to put pressure on those people (involved in animal testing) but we don't want them to think their children are going to be abducted. That's ridiculous. . . . The tone and tenor need to change, not just from a PR perspective but because we represent a noble cause. . . .
>
> The American public supports violence at times. But the animal rights movement is not at the stage now where violence or the rhetoric of violence is appropriate. . . . It's not even close.[43]

Some might suspect that Jonas's change in tenor was a play for a lighter sentence. I know Kevin Jonas personally, however, as an intelligent and sensitive young man, and certainly as somebody who would never publicly misrepresent his views on animals or activism for his own gain. While his earlier activism was motivated by justifiable outrage, he seems to have come to the conclusion that the expression of outrage must be focused and tempered in order to help the animals.

Jonas does not seem to be ruling out violence or the rhetoric of it completely. Rather he seems to be acknowledging the importance of public opinion—acknowledging that with a groundswell of public support, forcible acts against those who will not relinquish power can sometimes succeed, but that we must change hearts and minds in order to change the world. One might look at the activities of Sea Shepherd here: The

Sea Shepherd crew risk their own lives, and those of Japanese whalers, as they engage in dangerous missions on the high seas. While whaling nations try to portray the Sea Shepherd crew as terrorists, the badge doesn't stick with the public. Surely that is because the world loves whales and hates whalers. Public opinion is on Sea Shepherd's side, and public opinion matters.

The psychology of persuasion

Bizarro © Dan Piraro/King Features Syndicate

Public opinion is, of course, not the only reason to reject violence. Many people are attracted to the animal rights movement because of their basic rejection of violence—against any species, human or other. They know that one cannot build a compassionate world with bricks of hate.

I have been unwilling to condemn militant activism outright because none of us has a crystal ball—we cannot pretend to know the long-term effects of different types of activism. But watching the backlash against our movement in the UK provided me with some clarity. Robert Cialdini's book *Influence: The Psychology of Persuasion*[44] provided more.

Cialdini describes an experiment conducted by Jonathan Freedman: Freedman showed a group of boys an array of toys. He told them that one toy, the robot, was wrong to play with, and that if they played with it he would be angry and they would be punished. Then Freedman left the room and monitored the boys through one-way glass. Only one of the twenty-two boys played with the robot. Six weeks later, Freedman sent a colleague to the school to administer a drawing test. With no mention of the previous experiment, she took the boys to a room where there was an array of five toys, again including the robot. They were told they were welcome to play with the toys; 77 percent chose to play with the robot that had been forbidden to them earlier.

Then Freedman did a similar experiment with another group of boys. This time, however, on their first exposure to the robot, they were told that it was wrong to play with the robot but there was no threat to frighten the boys into obedience. Again, only one out of twenty-two boys touched the robot while Freedman was out of the room. The difference came six weeks later, when Freedman's colleague took the boys for the drawing experiment and told them they were welcome to play with the toys. This time only 33 percent chose the robot.

Cialdini explains that a strong threat will motivate immediate compliance but is unlikely to produce long-term commitment. He writes, "It seems clear that the threat had not taught the boys that operating the robot was wrong, only that it was unwise to do so when the possibility of punishment existed." He explains that the second group remembered that the reason they had not played with the robot was because they had made the decision that they did not want to, and they were likely to hold to that decision because humans tend toward consistency.

Similarly, militant activism can scare people for a while into doing the right thing by the animals. But unless we hope to run an animal-friendly oppressive regime, militancy seems unlikely to bring about the revolutionary societal shift in the treatment of other species we hope to see. Revolutions are only successful in the long term if they reflect the will of the people. That's why we must be profoundly concerned with public opinion and with doing our best to change not only laws but also hearts and minds. Educating people, and calling on their sense of kindness and justice, is the way to bring about lasting change.

"I've cared about animals my whole life. When I was seven, my brothers and some friends and I were the Frog Patrol. I grew up in Oklahoma, where, during summer nights, frogs would wander en masse onto the roads and get run over by cars. We'd stay up 'til four a.m. scooping them out of the street so they wouldn't get squished. . . .

"Everybody makes a personal choice to be vegetarian or not wear leather. I've gone back and forth, but like most people I aspire to be better. I'm not that radical, but I'm glad some people are—to drive the debate and keep the issues being discussed" (Grammy winner **Wayne Coyne** of the Flaming Lips).

PETA interview available online at www.peta2.com/OUTTHERE/o-FlamingLips.asp.

personal change

In the first chapter of this book I suggested that my goal was not to try to impose my values on the reader; rather, I hoped to get the reader to ask herself whether her behavior is in line with her own values. To that end, I have tried to bring to light some of the pervasive institutionalized cruelty that is hidden from society, which many of us might be unwittingly supporting.

Some people might have read this book and decided the one change they want to make is to stop frequenting entertainment events that use animals. As more parents refuse to bring their kids to animal circuses, eventually the market will cause them to be phased out. Some readers might change their wardrobe—and increase the demand for animal-friendly fabrics. Others might change their eating habits and commit to buying only free-range animal products. I hope many will now better understand why some of their friends are vegan, and decide to give that fun and cruelty-free lifestyle a go.

Flexitarian or Very Vegan?

It is important to remember that an all-or-nothing attitude does not help the animals. None of us will do all. Nobody reading this book is about to become a Jain, sweeping the path before himself so as not to harm insects. My friends in the vegan police remind me that I am not really vegan, and should call myself veganish, or "mostly vegan," because I choose to relax my standards both so that I don't go out of my mind, and because a more relaxed attitude seems to send the most helpful message.

Sometimes I actively choose food I know isn't vegan if I think that choice will have the greatest positive impact on the animals in the long run. If I am in a mainstream restaurant where I am surprised to find a veggie burger, I will try to order it, despite my general preference for salads, and despite my preference to avoid whatever traces of egg or cheese might be included in the burger if it is not advertised as vegan. At this stage of the game, it seems most important to make sure that the veggie burger sells well so that the restaurant won't take it off the menu. I might tell the waiter or manager how thrilled I am to find a veggie burger and that I would be even happier, and would come back often, if it were vegan. If the restaurateur sees enough demand for the veggie burger he will probably add more veggie options, and vegan options.

"I started my vegetarianism for health reasons, then it became a moral choice, and now it's just to annoy people."

Also, I like to dine with friends in nonvegan restaurants, and would never check whether or not the bread was made with some whey or honey, or whether the wine was made with a strictly vegan process. My hope is to bring the vegan lifestyle to the mainstream, and I think giving the impression that it is hard to be vegan could be counterproductive. It is not hard at all in modern cities if one is willing to relax the strictest standards in order to partake of modern life. Being willing to relax a little can make the difference between continuing to follow an animal-friendly lifestyle or not.

I have met people who tell me they went vegan for a while and then gave up because it was too hard. Now they eat absolutely anything—even bacon double cheeseburgers made with factory-farmed pork. That's crazy! It comes from the rigid idea that if one isn't totally vegan, one isn't helping at all, so one might as well just do nothing—and that isn't true. An article in the *New York Times* on the burgeoning business of cruelty-free fashion said that designers are aiming at potential customers identified in a survey by the consumer research company Mintel International as "occasional vegetarians."[45] The article said of those consumers, "They shop vegan selectively, as the Mintel study pointed out, but their 'purchasing power is paramount.'"

Every single compassionate choice makes a difference. It leads us closer to a world where making a compassionate choice is always an easy option because compassion has become the market leader.

Talk the walk

One of the most important things we can do is to be willing to raise our voices and say that what is happening to the animals matters. The power of just one voice can have enormous impact.

In one of her taped lectures, the spiritual teacher Marianne Williamson talks about an experiment done by NBC's *Dateline*.[46] The show had an actor pretending to be hurt and crying out for help. Nearby, two other actors were just hanging out and talking. In a candid camera–type of situation, *Dateline* watched the reactions of people walking by. Almost every person, as they saw two others ignore the cries for help, just kept walking. One person in twenty stopped, and then called for more help.

Here's the good part: Once one person stopped, every person who came into the situation afterward also stopped and was willing to get involved. As Marianne explains it, once one person acts from an awakened heart, others will follow. When one is willing to speak the word, others will listen. But we must be willing to take a stand and speak it in a way that others will hear. The problem is not that we don't have enough love; the problem is that we are whispering our love. Or we are yelling it in such a tone that it does not sound like love at all.

On that *Dateline* show, the one person in twenty was not whispering. Nor was she attacking or insulting those who had not stopped. That wasn't necessary. To completely change the situation, all it took was for one person to call out and say, "Somebody needs help over here."

Somebody does. The animals desperately need the help of those willing to call attention to their plight, to say that they matter, to speak the word. Don't whisper it—speak it loud. Speak it with laughter and love so others will be willing to listen. You cannot be shy as you speak it; your voice is crucial. You speak for those who cannot speak for themselves.

Mutts © Patrick McDonald/King Features Syndicate

Recommended Resource Groups

This is not a comprehensive list, but rather notes on some organizations with which I am involved and can vouch for:

PETA and PETA2

While **PETA.org** is packed with great information on all things animals, **PETA2** is particularly fun. It lets you know what all the hot musicians and actors are doing for animal rights and even gives you a chance to win tickets to their shows or other prizes. All that, while keeping you up to date on all the animal issues and what you can do to help.

Farm Sanctuary

A visit to an animal sanctuary can be a life-changing experience. **Farm Sanctuary** is the leader in the field. Founded by Gene Baur and Lorri Bauston back in the 1980s, the organization, now run by Baur, has glorious sanctuaries in beautiful Watkins Glen, New York, and in Orland, California, where you can get to know the animals and even stay overnight. Farm Sanctuary is also a leader on farm-animal protective legislation. Its many Web sites are packed with information, photographs, and footage. Learn more at www.FarmSanctuary.org.

other animal sanctuaries

Animal Acres is a warm and cozy farm animal sanctuary less than an hour from Los Angeles. It is run by Farm Sanctuary co-founder Lorri Bauston. It is a haven you must visit when you are in Los Angeles. You'll learn more on the Web at www.animalacres.org.

On a trip to Washington, D.C., I visited **Poplar Springs Animal Sanctuary** in Maryland and fell in love with Olivia the turkey. The Poplar Springs sanctuary is beautiful, and is exceptionally warm and friendly. Learn more at www.animalsanctuary.org.

The **Woodstock Farm Animal Sanctuary**, a newcomer on the scene, is like Farm Sanctuary's little sister. It is in the fun town of Woodstock about two and a half hours from New York City. Dan Piraro, whose *Bizarro* cartoons you saw all throughout this book, is a passionate supporter.

Canadians should learn about Gloria Grow's wonderful **Fauna Foundation** sanctuary near Montréal. Tom, whose tale is told in this book, and many other chimps are there. So are rescued farm animals. Check out www.faunafoundation.org/ff.

For dogs and cats, the premier sanctuary is **Best Friends** in Utah. Best Friends is a mighty rescue group that put forward a splendid effort during the Hurricane Katrina animal crisis. And Best Friends fought for and took in twenty-two of Vick's dogs. Already familiar with the group's great companion animal work, I became a firm supporter when I learned that I didn't have to worry about being served other animals at Best Friends fabulous fundraisers. (Check out the annual Hollywood Lint Roller Party!) While focusing on our companions, Best Friends is run by vegetarians and tries to be kind to everybody.

Legislation—HSUS

The **Humane Society of the United States (HSUS)** is an animal welfare rather than an animal rights organization, but since Wayne Pacelle came to the helm, we see some of its projects starting to reflect what many in society are coming to accept—that it is not really in the animals' welfare for them to be eaten or experimented on. The folks at HSUS won't shove veganism down your throat—though they might recommend some good plant-based recipes for you to try. Most notably, HSUS is the frontrunner on animal-friendly legislation. The HSUS Web site, www.hsus.org, and that for the Humane Society Legislative Fund, www.fund.org, and their associated alert lists are the best place to learn about animal-friendly voting or legislative bills, and what you can do to help them pass.

HSUS also has a fun Web site at www.humaneteen.org—the name says it all. Plus, the HSUS Hollywood Office puts on the annual Genesis Awards!

Health Issues

The **Physicians Committee for Responsible Medicine** is a superb place to learn about health issues—vegan diet and also about animal experimentation. Check out www.PCRM.org.

Vegan Outreach (www.VeganOutreach.org) specializes in vegan diet issues—a great resource.

Americans for Medical Advancement (www.CureDisease.com), provides comprehensive information on the human cost of animal experimentation.

animals in entertainment

SHARK (Showing Animals Kindness and Respect, www.SharkOnline.org), battles the abuse of animals at live events such as rodeos.

In Defense of Animals is a leader on zoos and circuses, and tackles a wide range of other issues, including fur, vivisection, and the dolphin slaughter. Learn more at www. IDAUSA.org.

conservation

The **Jane Goodall Institute** does wonderful conservation work, originally focused on chimps but now with wider coverage. It has a superb youth program, Roots and Shoots (www.RootsandShoots.org), with various memberships and discussion boards.

As its name suggests, **Sea Shepherd**, under the captainship of Paul Watson, guards the marine mammals. Learn more at www.SeaShepherd.org.

Ric O'Barry, who fights for the dolphins in Japan, is the marine mammal specialist for the **Earth Island Institute**. Find out more at www.earthisland.org/immp.

Grass Roots Groups

Two little engines that could are **Compassion Over Killing** (www.COK.net), a wonderfully effective group based in Washington, D.C., and **Mercy for Animals** (www. MercyForAnimals.org), founded in Ohio and now moving through Chicago and across the Midwest.

For Your Body and soul

Sting's wife, Trudi Styler, said at the New York **Jivamukti Yoga Studio** grand opening celebration, July 7, 2005—if you aren't taking loving-kindness out into the world, then you aren't a yogi, you are just a person with a mat. That's the Jivamukti philosophy.

What a joy it is to practice yoga in a haven where fur coats are not welcome, and where the café serves only vegan food. This hip center of the New York yoga scene is also a center for animal rights. Sharon Gannon and David Life, who founded Jivamukti Yoga—an international school—tell their teacher trainees to sign up for DawnWatch! You'll generally find that Jivamukti-trained teachers in your city will have animal-friendly leanings.

And finally, MEDIA!

Nobody in our society has more power than the media. For years, **DawnWatch** news alerts have been helping activists harness that power for the animals. DawnWatch.com is transitioning over to ThankingtheMonkey.com. Please visit **ThankingtheMonkey. com** to sign up for news alerts about media coverage of animal issues, and to learn what you can do to encourage more coverage.

Notes

CHAPTER 1 Welcome to the World of Animal Rights

1. George Thorndike Angell, *Autobiographical Sketches and Personal Recollections* (Boston: The American Humane Education Society, 1891).
2. Directed by Michael Apted, 2006.
3. In *Rattling the Cage* (Cambridge: Perseus, 2001), Wise argues for those rights for chimps, gorillas, and bonobos. He discusses the possibility of extending them to other animals in his subsequent book, *Drawing the Line* (Cambridge: Perseus, 2002).
4. As outlined in his book *Dominion: The Power of Man, the Suffering of Animals, and the Call to Mercy* (New York: St. Martin's Griffin, 2003).
5. Matthew Scully interview with Karen Dawn on KPFT Houston, April 28, 2003.
6. While I would love to claim this point as my own, David Stewart of Carolina Animal Action made the Amnesty analogy in a personal e-mail to me dated January 31, 2007.
7. Point made by Adam Weissman in personal e-mail dated December 22, 2006. He wishes to clarify that his comment was not intended as an outright rejection of the abolitionist perspective, but rather as a suggestion that the issue is more nuanced than some believe it to be.
8. Robert Cialdini, *Influence: The Psychology of Persuasion*, rev ed. (New York: Collins, 2007).
9. *Self*, July 2007, p. 34.
10. "Food Industry Must Coalesce or Lose War," *Feedstuffs*, April 2, 2007.
11. *Beyond Closed Doors* produced and directed by Hugh Dorigo, Sandgrain Films, 2006.
12. Ibid.
13. www.peta.org.
14. Said by Wyler at the closing of many Genesis Awards Ceremonies including one that aired on *Animal Planet*, May 6, 2006. Also noted in Carla Hall, "A Different Kind of Starring Role for Actress," *Los Angeles Times*, March 18, 2006, p. B1.
15. Adolf Hitler, *Mein Kampf*, James Murphy, trans. (Reedy, W.V.: Liberty Bell Publications, 2004), p. 134.
16. *New York Times*, "Corrections" box, March 15, 2005.
17. *Rivera Live,* hosted by Dan Abrams, July 20, 2000.
18. Peter Singer, *Animal Liberation* (New York: Random House, 1975).
19. Singer, *Practical Ethics*, 2nd ed. (Cambridge: Cambridge University Press, 1993).
20. Heard on Rush Limbaugh's radio show in 1992. He reiterates the thoughts in *The Way Things Ought to Be* (New York: Pocket Books, 1992), p. 108.
21. sci.rutgers.edu/forum/archive/index.php/t-72703.html.

CHAPTER 2 Slaves to Love: Pets

1. Opening Address, Taking Action for Animals, Washington, D.C., September 2, 2006.
2. Kenneth A. Gershman M.D., M.P.H., Jeffrey J. Sacks M.D., M.P.H., and John C. Wright Ph.D., "Which Dogs Bite? A Case-Control Study of Risk Factors." *Pediatrics* 93, no. 6 (June 1994), pp. 913–17.
3. Jon Mooallem, "The Modern Kennel Conundrum," *New York Times Magazine*, February 4, 2007, p. 42.
4. Charles Siebert, "New Tricks," *New York Times Magazine*, April 8, 2007.
5. "Brit's Pup Pick Leaves Animal Lovers Howling," *Boston Herald*, July 17, 2007.
6. "Paris Buys a Pup from Britney's Pet Store," WENN Entertainment News Wire Service, July 26, 2007.
7. *Dateline*, April 26, 2000.
8. Justin Gillis, "Cloned Food Products Near Reality," *Washington Post*, September 16, 2006.
9. Ruth La Ferla, "The Face of the Future," *New York Times*, December 15, 2005, p. G1.
10. www.atts.org/statistics.html.
11. Glen Bui, "Bite Prevention and Bite Statistics," cited in *Best Friends Animal Society*, 2005, American Canine Association, available at network.bestfriends.org/Library/Download.aspx.
12. Bill Torpy, Bill Rankin, "Vick Indicted: Feds: QB Involved in Dogfighting; Falcons Star Could Face Jail Time, Hefty Fine," *Atlanta Journal-Constitution*, July 18, 2007, p. 1A.
13. "Feds Search Vick's Property"; "'NFL—Second raid yields several more dog remains,' source says," *Seattle Times*, July 7, 2007.
14. Torpy and Rankin, "Vick Indicted: Feds: QB Involved in Dogfighting; Falcons Star Could Face Jail Time, Hefty Fine."
15. Bill Rankin and D. Orlando Ledbetter, "Vick's Lies Boosted Prison Time," *Atlanta Journal-Constitution*, Decmber 11, 2007, p. 1.
16. Available at www.youtube.com/watch?v=LSHQv9FhvVA.
17. Sam Farmer, "It Will Be Difficult to Find a Vick Jersey; Nike Suspends Sales of the Quarterback's Apparel and Shoes. Reebok Will Take Back Unsold Jerseys," *Los Angeles Times*, July 28, 2007, p. D1.
18. Mooallem, "The Modern Kennel Conundrum," *New York Times Magazine*, February 4, 2007, p. 42.
19. Jean Hofve, DVM, Animal Protection Institute, "Cosmetic Surgery for Dogs and Cats: Tail Docking, Ear Cropping, Debarking, Declawing"; available at www.idausa.org/facts/cossurgery.html.
20. Ibid.
21. "Oldest Dog Dies, Age 26," *Western Daily Standard*, April 1, 2003, p. 17.
22. "Pet Food Officer Sold Stock Before Recall," *Washington Post*, April 11, 2007, p. D03.
23. Christie Keith, "Your Whole Pet. Bigger Than You Think: The Story Behind the Pet Food Recall," *San Francisco Gate*, April 3, 2007.
24. "What's in That Food?" *Washington Post*, April 18, 2004, p. M7.
25. "Bye-Bye Birdie; Parrots' Demands Often Prompt Owners to Take Flight," *Washington Post*, August 27, 2001.

26. Culum Brown, "Not Just a Pretty Face," *New Scientist,* June 12, 2004.

27. Sarah Simpson, "Fishy Business," *Scientific American,* July 2001.

28. *Dateline,* September 18, 2005.

29. Keith, "Your Whole Pet. Bigger Than You Think: The Story Behind the Pet Food Recall."

30. "Pet Food Officer Sold Stock Before Recall," p. D03.

CHAPTER 3 All the World's a Cage: Animal Entertainment

1. Pliny the Elder, *Natural History: A Selection*, translated by John F. Healy (New York: Penguin, 1991), pp.108–13.

2. Mike Keele and Karen Lewis, "Asian Elephant North American Regional Studbook," Oregon Zoo, 2005.

3. Linda Goldston, "Suit Says Ringling Mistreats Elephants," *San Jose Mercury News,* August 18, 2000, p. 1B; and Jessica Graham, "Protesters to Greet Beastly Circus," *New York Post*, March 14, 2000, p. 3.

4. www.petatv.com/tvpopup/Prefs.asp?video=carson_barnes.

5. Bill Johnson, "Teen Tosses Animal-Abuse Claim into Wrong Ring," *Rocky Mountain News*, January 14, 2004, p. 7A.

6. "Ringling Brothers: Under Fire for Elephant Abuse," KTVU, September 3, 2004.

7. Rich Mckay, Sean Mussenden, "Activists Accuse Circus of Abuse; A New Report Repeats Allegations That Ringling Bros. Mistreats Elephants," *Orlando Sentinel*, September 25, 2003.

8. Fern Shen, "Three-Ring Bind; Some Say Circuses and Animals Don't Mix," *Washington Post*, April 11, 2002, p. C14.

9. Robert Meyerowitz, "Elephant in the Room," *Anchorage Press*, February 17, 2005.

10. A notable exception was Meyerowitz, "The Elephant in the Room."

11. Katie Pesznecker, "Maggie Graduates to Harness," *Anchorage Daily News*, May 21, 2007.

12. Kyle Hopkins, "Bob Barker Says Price Is Right to Move Maggie Off to California," *Anchorage Daily News*, August 28, 2007.

13. Tony Perry, "Elephant Behavior Studied in Hands-Off Training," *Los Angeles Times*, July 12, 2005, p. B5.

14. "Cruel and Usual," *U.S. News & World Report*, August 5, 2002.

15. Terri Langford, "With Gorilla on Run, Zoo, Police Clashed," *Dallas Morning News*, March 31, 2004.

16. Hilary Kindschuh, "Three Chimps Shot After Escape from Zoo Nebraska," *Lincoln Journal Star*, September 12, 2005.

17. Mark Branagan, "Chimp on Run Is Shot Dead," *Yorkshire Post*, December 10, 2005.

18. Olga Craig, "Chimp Tunnels to Freedom but Is Shot Dead by Zoo," *Sunday Telegraph*, September 30, 2007, p. 3.

19. Argo Films Limited Production for Thirteen/WNET New York and National Geographic Television in association with Trebitsch Produktion International. Aired on PBS November 19, 2000.

20. www.elephants.com/jenny/jenny_passes.htm.

21. Sally Kestin, "Not a Perfect Picture," *Sun-Sentinel*, May 16, 2004.

22. *Primetime*, October 27, 2005.

23. Boyd Harnell, "'Secret' Dolphin Slaughter Defies Protests," *Japan Times*, November 30, 2005.

24. Richard O'Barry and Keith Coulbourn, *Behind the Dolphin Smile* (Chapel Hill, N.C.: Algonquin Books of Chapel Hill, 1988).

25. www.chimpcollaboratory.org/news/movie.asp; film by Charles Spano.

26. Mary Braid, "Is Oliver the Chimp Half-Human?" *Daily Mail* (London), March 1, 2003.

27. James Shreeve, "Oliver's Travels," *Atlantic Monthly*, October 2003.

28. Lisa Sandberg, "Chimp Sanctuary Woes Existed Long before Court Order," *Houston Chronicle*, November 26, 2006, p. 4.

29. Ibid.

30. Dana Bartholomew, "Hollywood Under Fire in Death of 2nd Horse," *Los Angeles Daily News*, April 28, 2005, p. N1.

31. Part of the *Watchdog* series, hosted by Karen Dawn on KPFK Radio, April 19, 2004; archives can be heard online at www.WatchdogRadio.com.

32. Dolphin Project video, produced by Diana Thater, 2000.

33. Wendy-Anne Thompson, "Animal-Rights Protesters Try to Drum Up Support," *Calgary Herald*, July 10, 2003.

34. Andrea Elliot, "Bringing Down the Bulls," *Miami Herald*, January 22, 2003, p.2.

35. *Daily Show with Jon Stewart*, 2006; segment online at www.comedycentral.com/sitewide/media_player/play.jhtml?itemId=75362.

36. Kenneth R. Weiss, "Sportfishing Blamed in Depletion," *Los Angeles Times*, August 27, 2004, p. B1.

37. George Tasker, "Catch and Release Also a Problem," *Miami Herald*, October 16, 2005.

38. Maryland Department of Natural Resources Web site, "DNR Reminds Anglers to Use Extra Care in Summer Catch-and-Release Fishing," August 3, 2000; available at dnrweb.dnr.state.md.us/dnrnews/pressrelease2000/080400.html.

39. Jerry Adler, "Can Barbaro Beat the Odds?" *Newsweek*, June 5, 2006.

40. *Larry King Live*, January 30, 2007.

41. Richard Rosenblatt, Associated Press, "Barbaro's Injury Focuses Interest on a Safer Way to Race," June 9, 2006.

42. Rich Hofmann, "Racing Brings Up the Rear in Safety," *Philadelphia Daily News*, May 23, 2006.

43. "Report: Failed Drug Tests in Horses Increasing," *USA Today*, December 11, 2006.

44. On KPFK's *Watchdog*, May 17, 2004; archives online at www.WatchdogRadio.com.

45. "Kentucky Considers Medication Limits for Racehorses," reported by Adam Hochberg, *All Things Considered*, January 31, 2005.

46. William C. Rhoden, "An Unknown Filly Dies, and the Crowd Just Shrugs," *New York Times*, May 25, 2003.

47. *Real Sports*, November 23, 2004.

48. "A Dog's Life Ain't What It Used to Be," *Independent*, January 17, 2005.

49. Sandra Gonzales, "The Race Is On—for Giant Turtles," *San Jose Mercury News*, April 15, 2007.
50. *The Colbert Report*, April 12, 2007.
51. *Politically Incorrect*, March 20, 2001.

CHAPTER 4 Fashion Victims: Animal Clothing

1. "The Lamb on the Runway," *New York Times,* August 11, 2005, p.61.
2. Rebecca Aldworth, "Make This Year's Seal Hunt the Last," *Christian Science Monitor,* March 18, 2005.
3. *Larry King Live*, December 11, 2005.
4. Reported by Patrick McGrath on Fox 5, Washington, D.C., February 10, 2007.
5. Tina Cassidy, "With Faux Looking So Real These Days, You Can't Tell Who's . . . Faking It," *Boston Globe*, January 12, 2006.
6. Footage online at www.petaindia.com/cleath.html.
7. Richard E. Sclove et al., *Community-Based Research in the United States* (Amherst, Mass.: The Loka Institute, 1998).
8. Edith Stanley, "Chicken Again? These Gators Get a Steady Diet of Dead Fowl," *Los Angeles Times*, June 10, 2001.
9. "How About a Little 'Rigoletto'?" *Christian Science Monitor*, January 30, 2006; compiled from wire reports by staff.
10. "A Shipload of Sheep Without a Harbor," *Los Angeles Times*, October 12, 2005.
11. Michael Weiss, "Animal Einsteins," *Reader's Digest*, November 2006.
12. Charles Siebert, "The Animal Self," *New York Times*, January 22, 2006.
13. Asian News International, "Spiders Love to Snuggle Up," published in the *Hindustan Times*, March 20, 2007.
14. Hidehiro Watanabe, Makoto Mizunami, "Pavlov's Cockroach: Classical Conditioning of Salivation in an Insect," June 13, 2007, available on PloS One at www.plosone.org/article/fetchArticle.action?articleURI=info:doi/10.1371/journal.pone.0000529.
15. Ruth La Ferla, "Uncruel Beauty," *New York Times*, January 11, 2007.
16. Scully, *Dominion*, pp. 42–43.

CHAPTER 5 Deconstructing Dinner

1. Sondra Crosby, "Lessons From My Pig Winnie," *Boston Globe*, March 19, 2005.
2. Joby Warrick, "They Die Piece by Piece," *Washington Post*, April 10, 2001.
3. Peter Singer and Jim Mason, *The Way We Eat: Why Our Food Choices Matter* (New York: Rodale, 2006), p. 67.
4. Eric Schlosser, *Fast Food Nation* (New York: HarperPerennial, 2002), p. 171.
5. Michael Pollan, "This Steer's Life," *New York Times Magazine*, March 31, 2002.
6. Freedom of Information Act, #94-363, "Poultry Slaughtered, Condemned, and Cadavers," June 30, 1994, cited in Karen Davis, Ph.D., *Prisoned Chickens Poisoned Eggs: An Inside Look at the Modern Poultry Industry* (Summertown, Tenn.: Book Publishing Company, 1996).
7. Daniel Zwerdling, "A View to Kill," *Gourmet*, June 2007, p. 94.

8. CNN *Money*, July 20, 2004; see money.cnn.com/2004/07/20/news/fortune500/kfc/.

9. Conducted in April and July 2006 and reported on www.ButterballCruelty.com.

10. Fox Cleveland I-team Investigation, July 25, 2007, video footage available at www .mercyforanimals.org/HOR/.

11. "A Halal Slaughterhouse Provides Nourishment for a Far-Flung Culture," *New York Times*, March 9, 2005, p. B1.

12. "Inquiry Finds Lax Federal Inspections at Kosher Meat Plant," *New York Times*, March 10, 2006, p. A16.

13. *60 Minutes*, "Big Chicken," December 19, 1999.

14. "USDA Says Rule on Livestock Stops Applies to Trucks," *Des Moines Register*, September 29, 2006.

15. Zwerdling, "A View to Kill."

16. Stephanie Brown and John Youngman, "End the Rough Ride for Farm Animals," *Globe and Mail*, March 15, 2006.

17. John McGlone, "Fatigued Pigs: The Final Link," *Pork*, March 2006, pp. 14–16; and P. D. Warris et al., "Longer Journeys to Processing Plants Are Associated with Higher Mortality in Broiler Chickens," *British Poultry Science* 33, no. 1 (March 1992), pp. 201–06.

18. "Research Looks at Transport Losses," *Feedstuffs*, April 16, 2006.

19. www.youtube.com/watch?v=DVFB226416s.

20. Marc Kaufman, "In Pig Farming, Growing Concern: Raising Sows in Crates Is Questioned," *Washington Post*, June 19, 2001.

21. Lauren Etter, "Smithfield to Phase Out Crates. Big Pork Producer Yields to Activists, Customers, on Animal-Welfare Issue," *Wall Street Journal*, January 25, 2007, p. 14.

22. Jeff Tietz, "Bosshog," *Rolling Stone*, December 14, 2006, p.89.

23. Zwerdling, "A View to Kill."

24. G. T. Tabler and A. M. Mendenhall, "Broiler Nutrition, Feed Intake and Grower Economics," *Avian Advice* 5, no. 4 (Winter 2003), p. 9, in Singer and Mason, *The Way We Eat*.

25. M. O. North and D. D. Bell, *Commercial Chicken Production Manual*, 4th ed. (New York: Van Nostrand Reinhold, 1990), p. 456, in Singer and Mason, *The Way We Eat*.

26. Ron Reagan, MSNBC, March 30, 2005.

27. Dan Noyes, "I-Team" Investigative Report, ABC-7 San Francisco, September 16, 2003.

28. Available at www.ctv.ca/servlet/ArticleNews/story/CTVNews/20070710/foie_gras _070710/20070710/ and tinyurl.com/2me7wd.

29. Pollan, "This Steer's Life."

30. F. M. Mitlöhner, M. L. Galyean, and J. J. McGlone, "Shade Effects on Performance, Carcass Traits, Physiology, and Behavior of Heat-stressed Feedlot Heifers," *Journal of Animal Science* 80 (2002), pp. 2043–50, in Singer and Mason, *The Way We Eat*.

31. Singer and Mason, *The Way We Eat*.

32. GAO Report to the Secretary of the Interior, "Improvements Needed in Federal Wild Horse Program," August 1990.

33. *The Misfits*, directed by John Huston, written by Arthur Miller, 1961.

34. Peter Singer, *Animal Liberation* (New York: New York Review, 1975), p. 184.

35. Lynne Sneddon, V. A. Braithwaite, and M. J. Gentle, "Do Fish Have Nociceptors? Evidence for the Evolution of a Vertebrate Sensory System," *Proceedings of the Royal Society* 270,

no. 1520 (2003), pp. 1115–21, in Victoria Braithwaite, "Hooked on a Myth," *Los Angeles Times*, October 8, 2006.

36. Ken Olbermann, *Countdown*, April 30, 2003.

37. Lori Valigra, "Under the Sea—Sotto Voce No More," *Christian Science Monitor*, November 8, 2001, p. 16.

38. Ibid.

39. Robert Matthews, "Fast-Learning Fish Have Memories That Put Their Owners to Shame," *Telegraph*, March 10, 2004.

40. "Fish Story: Mad Scientists Prove Pet Fish Have More on the Ball Than We Thought," *Pittsburgh Post-Gazette*, January 24, 2006.

41. Letters, *Reader's Digest*, May 4, 2006.

42. Singer and Mason, *The Way We Eat*, p. 132.

43. Ibid.

44. "Seal's Death Shows Need for Gillnet Rules," *Star Bulletin*, October 24, 2006.

45. "Mating Season Can Be Perilous for Right Whales," *Washington Post*, November 25, 2006, p. A02.

46. Tim Weiner, "In Mexico, Greed Kills Fish by the Seaful," *New York Times*, April 10, 2002.

47. Steve Gibson. "Netters Discarding Dead Mullet,"*Herald Tribune*, April 8, 2002.

48. Erik Stokstad, "Global Loss of Biodiversity Harming Ocean Bounty," *Science* 314, no. 5800 (November 3, 2006), p. 745.

49. "War at Sea: Greenpeace Fights to Save Dolphins from the Nets," *Independent*, March 26, 2005.

50. Kenneth R. Weiss, "Fish Farms Become Feedlots of the Sea," *Los Angeles Times*, December 9, 2002, p. A1.

51. Alexandra Morton, in Weiss, "Fish Farms."

52. *Nightline*, November 28, 2006.

53. Marian Burros, "Stores Say Wild Salmon, but Tests Say Farm Bred," *New York Times*, April 10, 2005.

54. "Colombian Navy Trains Its Sights on Shark Hunters," *Financial Times*, October 24, 2006.

55. "Hidden Cost of Shark Fin Soup: Its Source May Vanish," *New York Times*, January 5, 2006.

56. Temple Grandin, quoted in Scully, *Dominion*, p. 245.

57. Pollan, "This Steer's Life."

58. Charlotte Ferrell Smith, "Evaporated Milk: Cow Vanishes from New Owner's Farm, Embarks on 20-mile Journey to Be Reunited with Her Calf," *Charleston Daily Mail*, September 20, 2005, p. P1A.

59. S. Waage et al., "Identification of Risk Factors for Clinical Mastitis in Dairy Heifers," *National Veterinary Institute*; available at jds.fass.org/cgi/reprint/81/5/1275.pdf.

60. Verlyn Klinkenborg, "Holstein Dairy Cows and the Inefficient Efficiencies of Modern Farming," *New York Times*, January 5, 2004.

61. Alex Pulaski and Andy Dworkin, "Mad Cow Disease Raises Questions About Hodgepodge That Is Hamburger," Newhouse News Service, February 19, 2004.

62. Jim Bodor and Jacqueline Reis, "How Safe Is Our Beef? Inspectors, Farmers, Brokers Part of Mad Cow Defense," *Sunday Telegram* (Massachusetts), January 18, 2004.

63. Douglas Fischer, "Meat Eaters Getting Curious: State's Supply Deemed Safe, but Local Consumers Face a Tough Time Tracking Hamburgers' History," *Oakland Tribune*, January 1, 2004.

64. *Stall Street Journal*, April 1973, put out by Provimi Inc.

65. *Baraka*, experimental documentary film directed by Ron Fricke, 1992.

66. Natalie Jordi, February 2007; online at www.plentymag.com/blogs/ecoeats/2007/02/wait_nowe_actually_meant_how_t.php.

67. Tobias Young, "Recycling Chickens: Farmers Turn to Composting Amid Collapsed Spent-hen Market," Santa Rosa *Press Democrat*, November 22, 2006.

68. Peter Singer and Karen Dawn, "Back at the Ranch, a Horror Story," *Los Angeles Times*, December 1, 2003.

69. "What Is Organic? Powerful Players Want a Say," *New York Times*, November 1, 2005, p. C1.

70. Ibid.

71. "Organic: Critics Say Dairy Tests the Boundaries and Spirit of What 'Organic' Means," *Chicago Tribune*, August 20, 2006.

72. Andrew Martin, "Meat Labels Hope to Lure the Sensitive Carnivore," *New York Times*, October 24, 2006, p. 1A.

73. Sean Callebs, CNN, July 25, 2004.

74. Jennifer Wolcott, "'Cage-free' Eggs: Not All They're Cracked Up to Be?"; available at www.csmonitor.com/2004/1027/p15s02-lifo.html.

75. "Advocates Challenge Humane-Care Label on Md. Eggs: Birds Are Cruelly Caged, Lawsuit Argues," *Washington Post*, September 19, 2005.

76. Steven M. Wise, *Drawing the Line: Science and the Case for Animal Rights* (Cambridge, Mass.: Perseus, 2002).

77. Ibid.

78. Weiss, "Animal Einsteins."

79. Alexei Barrionuevo, "Bees Vanish; Scientists Race for Reasons," *New York Times*, April 24, 2007, p. F1.

80. *CBS 60 Minutes*, "What's with the Bees?" October 30, 2007.

81. Michael Pollan, *The Omnivore's Dilemma* (New York: Penguin, 2006), p. 362.

82. Spencer Vaa, "Reducing Wounding Losses," South Dakota Department of Game, Fish, and Parks, 2004; available at www.sdgfp.info/Wildlife/hunting/waterfowl/WoundingLosses.htm.

83. Raizel Robin, "The 5th Annual Year in Ideas"—"In Vitro Meat," *New York Times*, December 11, 2005.

84. Traci Hukill, "Controversy," *Chicago Sun Times*, July 16, 2006, p. B05.

85. Ibid.

86. "When Meat Is Not Murder: Would You Eat This Steak If It Had Been Grown in a Petri Dish?" *Guardian*, August 13, 2005.

87. Steve Fishman, "The Diet Martyr," *New York*, March 15, 2004.

88. Donald Watson obituary, *The Times*, December 8, 2005.

Notes

89. *Fast Food Nation*, directed by Richard Linklater, Fox Searchlight, 2006.

90. Schlosser, *Fast Food Nation*, p. 197.

91. Nina Planck, "Leafy Green Sewage," *New York Times*, September 21, 2006.

92. Schlosser, *Fast Food Nation*, p. 206.

93. Ibid.

94. Rick Weiss, "FDA Is Urged to Ban Carbon-Monoxide-Treated Meat," *Washington Post*, February 20, 2006, p. A01.

95. Rick Weiss, "Studies Attest to Buyers' Focus on Color of Meat," *Washington Post*, February 22, 2006, p. A07.

96. Sandi Doughton, "Number of Mad-Cow Tests in NW Didn't Reach Federal Agency's Goal," *Seattle Times*, February 24, 2004.

97. Editorial Desk, "More Mad Cow Mischief," *New York Times*, May 8, 2004, p. 16.

98. Guy McKhann et al., *Alzheimer Disease and Associated Disorders* 2 (1989), pp. 100–109.

99. Michael Greger, M.D., "Could Mad Cow Disease Already Be Killing Thousands of Americans Every Year?" CommonDreams.org, January 7, 2004.

100. Rick Weiss, "FDA Rules Override Warnings About Drug—Cattle Antibiotic Moves Forward Despite Fears of Human Risk," *Washington Post*, March 4, 2007, p. A01.

101. *Everwood*, May 19, 2003.

102. "Dirty Birds," *Consumer Reports*, January 2007.

103. Schlosser, *Fast Food Nation*, p. 140.

104. R. Sinha, N. Rothman, E. D. Brown, C. P. Salmon, M. G. Knize, C. A. Swanson, S. C. Rossi, S. D. Mark, O. A. Levander, and J. S. Felton, "High Concentrations of the Carcinogen 2-amino-1-methyl-6-phenylimidazo-[4,5-b]pyridine (PhIP) Occur in Chicken but Are Dependent on the Cooking Method," *Cancer Research* 55, no. 20 (1995), pp. 4516–19, in "Heterocyclic Amines in Cooked Meats"; available online at cis.nci.nih.gov/fact/3_25.htm.

105. Marian Burros, "Chicken With Arsenic? Is That O.K.?" *New York Times*, April 4, 2006, p. F6.

106. Matt McKinney, "Sampling Finds Arsenic in Most Chicken; A Locally Produced Study of Chicken Found That Most Sold at Supermarkets and All Sold at Fast-Food Restaurants Contained Arsenic," *Minneapolis Star Tribune*, April 6, 2006, p. 1D.

107. Michael Hawthorne and Sam Roe, "U.S. Safety Net in Tatters," *Chicago Tribune*, December 12, 2005.

108. Sam Roe and Michael Hawthorne, "Toxic Risk on Your Plate," *Chicago Tribune*, December 11, 2005.

109. Hawthorne and Roe, "U.S. Safety Net in Tatters."

110. Ibid.

111. Peter Waldman, "Fish Line. Mercury and Tuna: U.S. Advice Leaves Lots of Questions," *Wall Street Journal*, August 2, 2005.

112. Marion Nestle, *What to Eat* (New York: Northpoint Press, 2006), p. 191.

113. Hawthorne and Roe, "U.S. Safety Net in Tatters."

114. *Heart Disease Weekly*, January 19, 2002.

115. Marian Burros, "Industry Money Fans Debate on Fish," *New York Times*, October 17, 2007, p. F5.

116. *Science*, January 9, 2006.
117. David Biello, "Bringing Cancer to the Dinner Table: Breast Cancer Cells Grow Under Influence of Fish Flesh," "Rachel's Democracy & Health News," no. 903, April 19, 2007, available online at www.precaution.org/lib/07/prn_dhn070419.htm.
118. Charlie Fidelman, "Research Suggests Dairy Can Be Deadly. Ovarian Cancer Risk; Farmers Group Dismisses Study," *Montreal Gazette*, August 11, 2005, citing *Journal of Cancer*, August 11, 2005.
119. P. Bertron, N. D. Barnard, and M. Mills, "Racial Bias in Federal Nutrition Policy, Part I: The Public Health Implications of Variations in Lactase Persistence," *Journal of the National Medical Association* 91 (1999), pp. 151–57.
120. Reporters Blow Whistle on Fox News, www.youtube.com/watch?v=axU9ngbTxKw.
121. "Managing Editors: Developments in the News Industry for Feb. 10–18," Associated Press, February 18, 2003.
122. "Reporter's Jury Award Tossed Out," *St. Petersburg Times*, February 15, 2003.
123. Reporters Blow Whistle on Fox News.
124. Benjamin Spock, M.D., and Steven J. Parker, M.D., *Dr. Spock's Baby and Child Care*, rev. ed. (New York: Pocket Books, 1998), p. 331.
125. Ibid., p. 332.
126. Ibid., p. 199.
127. Ibid., p. 342.
128. Nina Planck, "Death by Veganism," *New York Times*, May 21, 2007.
129. Clark Hoyt, "The Danger of the One-Sided Debate," *New York Times*, June 24, 2007.
130. "Position of the American Dietetic Association and Dietitians of Canada: Vegetarian Diets," *Journal of the American Dietetic Association* 103, no. 6 (June 2003), pp. 748–65.
131. B. J. Trock et al., "Meta-analysis of Soy Intake and Breast Cancer Risk," *Journal of the National Cancer Institute* 98, no. 7 (April 5, 2006), pp. 459–71.
132. *American Journal of Epidemiology*, January 8, 2007; available at tinyurl.com/2ckbgb. Those who consumed 3 milligrams of isoflavones (a phytoestrogen found in soy foods) per day had a 44 percent lower risk than women who consumed less than 1 milligram. There was no significant evidence linked to any other foods or nutrients with ovarian cancer risk.
133. *International Journal of Cancer* (April 15, 2007), reported in "Animal Protein & Fat Raise Endometrial Cancer Risk," *Scientific American*, March 21, 2007; available at www.sciam.com/article.cfm?alias=animal-protein-amp-fat-ra&chanId=sa003.
134. E. F. Taylor, V. J. Burley, D. C. Greenwood, and J. E. Cade, "Meat Consumption and Risk of Breast Cancer in the UK Women's Cohort Study," *British Journal of Cancer*, no. 96, 2007, pp. 1139–1146.
135. "Red Meat Linked to Breast Cancer," *Age*, April 4, 2007.
136. Timothy J. Key et al., "Mortality in Vegetarians and Nonvegetarians: Detailed Findings from a Collaborative Analysis of 5 Prospective Studies," *American Journal of Clinical Nutrition* 70, no. 3 (September 1999), pp. 516S–24S.
137. Catherine R. Gale et al., "IQ in Childhood and Vegetarianism in Adulthood," *British Medical Journal* (December 15, 2006).
138. Monika Grillenberger et al., "Animal Source Foods to Improve Micronutrient Nutri-

tion in Developing Countries," *Journal of Nutrition* 133 (November 2003), pp. 3957S–64S.

139. Jonathan H. Siekmann et al., "Kenyan School Children Have Multiple Micronutrient Deficiencies, but Increased Plasma Vitamin B-12 Is the Only Detectable Micronutrient Response to Meat or Milk Supplementation," *Journal of Nutrition* 133 (November 2003), pp. 3972S–80S.

140. Harvey Diamond, quoted in Gail Davis, *Vegetarian Food for Thought* (Troutdale, Ore.: NewSage Press, 1999).

141. Carl Lewis, in Jannequin Bennet, *Very Vegetarian* (Nashville, Tenn.: Thomas Nelson, 2001).

CHAPTER 6 Animals Anonymous: Animal Testing

1. Carina Dennis, "Cancer: Off by a Whisker," *Nature* 442 (August 17, 2006), pp. 739–41.

2. Gardiner Harris, "New Drug Points Up Problems in Developing Cancer Cures," *New York Times*, December 21, 2005.

3. Dennis, "Cancer: Off by a Whisker."

4. C. Ray Greek, M.D., and Jean Swingle Greek, DVM, *Sacred Cows and Golden Geese: The Human Cost of Experiments on Animals* (New York: Continuum, 2001).

5. "Gender and Medicine: Some Drugs Affect Men and Women Differently," *World News Tonight*, January 17, 2006.

6. Interview with Karen Dawn on KPFK's *Watchdog* series, March 29, 2004; available at www.WatchdogRadio.com.

7. "The Thalidomide Disaster," *Time*, August 10, 1962.

8. *Experimental Molecular Pathology Supplement* 2 (1963), pp. 81–106; *Federation Proceedings* 26 (1967), pp. 1131–36; and *Teratogenesis, Carcinogenesis, and Mutagenesis* 2 (1982), pp. 361–74.

9. Deborah Wilson, "Testing Drugs on Animals No Longer Suitable Option," *Arizona Republic*, April 23, 2006.

10. "Lab Clue into Catastrophic Drug Trial," *New Scientist* 2582 (December 16, 2006), p. 6.

11. Jerry Avorn, on "Bitter Medicine: Pills, Profit and the Public Health," ABC, May 29, 2002.

12. Advertising Standards Authority, "Non-broadcast Adjudications," Association of Medical Research Charities, October 5, 2005; available at www.asa.org.uk/asa/adjudications/non_broadcast/Adjudication+Details.htm?adjudication_id=40340.

13. *Frontline*, March 27 and April 3, 2001; available at www.pbs.org/wgbh/pages/frontline/shows/organfarm/.

14. Ibid.

15. Ibid.

16. Dr. Sharon Levine, Kaiser Permanente Medical Group, on "Bitter Medicine."

17. Marcia Angell, "The Truth About the Drug Companies," *New York Review of Books* 51 (July 15, 2004).

18. Peter Jennings, on "Bitter Medicine."

19. Unidentified woman representative of Group Health on "Bitter Medicine."

20. Rita Rubin, "Vioxx, Celebrex Overprescribed?" *USA Today*, January 23, 2005.

21. Robert Goldberg, "Bitter Pill: The Truth About Drug Spending," *National Review*, June 10, 2002.
22. Angell, "The Truth About the Drug Companies."
23. "Bitter Medicine."
24. Angell, "The Truth About the Drug Companies."
25. Deborah Blum, *The Monkey Wars* (New York: Oxford University Press, 1994).
26. Harry Harlow, in a 1974 interview in *Pittsburgh Press Roto*, quoted in Blum, *The Monkey Wars*, p. 92.
27. Blum, *The Monkey Wars*, p. 97.
28. John P. Capitanio, Sally P. Mendoza, Nicholas W. Lerche, and William A. Mason, "Social Stress Results in Altered Glucocorticoid Regulation and Shorter Survival in Simian Acquired Immune Deficiency Syndrome," *Proceedings of the National Academy of Sciences of the United States of America* 95, no. 8 (April 14, 1998), pp. 4714–19.
29. whitecoatwelfare.org/cameron.shtml.
30. "Primates, Orphans Give Clue to Affects of Childhood Trauma," Associated Press, November 25, 2005.
31. Erica Goode, "Nerve Damage to Brain Linked to Heavy Use of Ecstasy Drug," *New York Times*, October 30, 1998.
32. George A. Ricaurte, Jie Yuan, George Hatzidimitriou, Branden J. Cord, and Una D. McCann, "Severe Dopaminergic Neurotoxicity in Primates After a Common Recreational Dose Regimen of MDMA 'Ecstasy,'" *Science* 297 (September 27, 2002), p. 2260.
33. Donald G. McNeil Jr., "Report of Ecstasy Drug's Great Risks Is Retracted," *New York Times*, September 6, 2003, p. A8.
34. Peter Wilson, "Animals' True Nature Will Out," *Weekend Australian*, November 18, 2006, p. 25.
35. Project B (Roselli), "A Ram Model of Neurendocrine Function," OHSU Protocol number A150, November 17, 2003.
36. John Schwartz, "Of Gay Sheep, Modern Science, and the Perils of Bad Publicity," *New York Times*, January 25, 2007.
37. Wilson, "Animals' True Nature Will Out."
38. Ibid.
39. "Vitamin C Might Cut Some Risks of Pregnant Smoking," Associated Press State & Local Wire, May 2, 2005.
40. Letter available at femfatalities.com.
41. Stephanie Saul, "Drug Makers Race to Cash In on Fight Against Fat," *New York Times*, April 3, 2005.
42. Rob Stein, "Obesity Passing Smoking as Top Avoidable Cause of Death," *Washington Post*, March 10, 2004.
43. P. K. Newby, Katherine L. Tucker, and Alicja Wolk, "Risk of Overweight and Obesity Among Semivegetarian, Lactovegetarian, and Vegan Women," *American Journal of Clinical Nutrition* 81, no. 6 (June 2005), pp. 1267–74.
44. Saul, "Drug Makers Race to Cash In on Fight Against Fat."
45. Johanna Neuman, "Obesity Fuels Their Fervor; Three Well-Known Nutrition Activists Take Business, Science, the Government and Us to Task," *Los Angeles Times*, July 26, 2004.

Notes

46. archives.cnn.com/2002/US/08/18/terror.tape.main/index.html and archives.cnn .com/2002/US/08/19/terror.tape.chemical/index.html.

47. Marie Woolf, "Military Lab Tests on Live Animals Double in Five Years," *Independent on Sunday*, May 14, 2006.

48. ABC-7 Denver, April 7 2006; available at www.thedenverchannel.com/news/8507368/ detail.html.

49. www.youtube.com/watch?v=bCdKIZRz_WM.

50. C. J. Chivers, "Tending a Fallen Marine, With Skill, Prayer and Fury," *New York Times*, November 2, 2006.

51. Kirk Loggins, "Report of Dogs Being Used to Test Electric Chair a Hoax," *Tennessean*, November 5, 1999.

52. Richard Willing, "Critics Condemn Electric Chair," *USA Today*, September 7, 2001, p. 3A.

53. Ray Cooklis, "The Chair," *Cincinnati Inquirer*, September 2, 2001.

54. Thomas Hartung and Alan Goldberg, "Protecting More than Animals: Reducing Animal Suffering also Has the Added Benefit of Yielding More Rigorous Safety Tests," *Scientific American*, January 2006, pp. 84–91.

55. *Boston Legal*, January 30, 2007.

56. Hartung and Goldberg, "Protecting More than Animals."

57. Zeeya Merali Lyon, "Human Skin to Replace Animal Tests," NewScientist.com, July 25, 2007.

58. B. Ekwall, "Overview of the Final MEIC Results: II. The In Vitro-In Vivo Evaluation, Including the Selection of a Practical Battery of Cell Tests for Prediction of Acute Lethal Blood Concentrations in Humans," *Toxicology in Vitro* 13 (1999), pp. 665–73.

59. Rokke reports on eight months of work at HLS in 1996–97 in *Behind the Mask*, a documentary by Shannon Keith, 2006. The film's first award was Best Documentary at The Other Venice Film Festival, 2007.

60. "Questions and Answers about the Animal Welfare Act and Its Regulations for Biomedical Research Institutions"; available at www.nal.usda.gov/awic/legislat/regsqa.htm.

61. "God (or Not), Physics and, of Course, Love: Scientists Take a Leap," *New York Times*, January 4, 2005.

62. "Emotional Rats," *Time*, April 10, 1939.

63. "Rats Capable of Reflection on Mental Processes," *Science Daily*, March 9, 2007, available online at www.sciencedaily.com/releases/2007/03/070308121856.htm.

64. Available at www.sciencedaily.com/releases/2007/03/070308121856.htm.

65. Peter Gorner, "Animals Enjoy Good Laugh Too, Scientists Say," *Chicago Tribune*, April 1, 2005, p. 11.

66. Natalie Angier, "Smart, Curious, Ticklish. Rats?" *New York Times*, July 24, 2007.

67. Nicholas Bakalar, "Rat to Rat, Kindness Takes Hold," *New York Times*, July 10, 2007.

68. Judah Folkman, quoted in Lawrence Osborne, "Fuzzy Little Test Tubes," *New York Times*, July 30, 2000.

69. Osborne, "Fuzzy Little Test Tubes."

70. "PETA Says Tape Shows Rat Research Violations," *Washington Post*, April 19, 2002.

71. www.neavs.org/programs/brochures/brochures_rat.htm.

72. Osborne, "Fuzzy Little Test Tubes."

73. Gina Kolata, "Of Mice and Men: Why Test Animals to Cure Human Depression?" *New York Times*, March 28, 2004.

74. P. Pound et al., "Where Is the Evidence That Animal Research Benefits Humans?" *British Medical Journal* 328 (2004), pp. 514–17.

75. Adam Wishart, *Animal Testing: Monkeys, Rats and Me*, original airdate November 30, 2006, BBC2.

76. Ibid.

77. Adam Wishart, "What Felix the Monkey Taught Me About Animal Research," *Mail on Sunday*, November 26, 2006.

78. Wishart, *Animal Testing*.

79. Peter Singer, "Setting Limits on Animal Testing," *Sunday Times* (London), December 3, 2006, p. 20.

80. Personal e-mail correspondence, December 26, 2006.

81. Cabeza de Vaca's *Adventures in the Unknown Interior of America* [1542], copyright ©1961 by The Crowell-Collier Publishing Company [but not renewed]. Reprinted 1983 by University of New Mexico Press.

82. Interview of Charles Hall by Michael E. Salla, Ph.D., reported on www.alienshift.com.

83. I noted in chapter 2 that I judge both specious and speciesist those arguments that suggest normal human lives are worth more than other animal lives not because we belong to the human species, but rather because of attributes that are most apparent in normal humans.

84. *Inside the Actors Studio*, Episode #12.4, original airdate October 30, 2005.

85. Wishart, *Animal Testing*. 2006.

86. Leslie Wolfson et al., "Dopamine Mechanisms in the Subthalamic Nucleus and Possible Relations to Hemiballism and Other Movement Disorders," in Arnold J. Friedhoff and Thomas N. Chase, eds., *Gilles de la Tourette Syndrome* (New York: Raven Press, 1982).

87. www.newhopeforparkinsons.com/web/pid/61/.

88. EC/IC Bypass Study Group, "Failure of Extracranial-intracranial Arterial Bypass to Reduce the Risk of Ischemic Stroke. Results of an International Randomized Trial," *New England Journal of Medicine* 313, no. 19 (1985), pp. 1191–1200.

89. Alok Jha and Paul Lewis, "Scientist Backs Animal Testing for Cosmetics," *Guardian*, March 4, 2006.

90. www.speakcampaigns.org/felix/rip.php.

91. *Chimpanzees: An Unnatural History*, produced by Allison Argo, PBS *Nature* series, original airdate November 5, 2006.

92. Will Dunham, "U.S. Stops Breeding Chimps for Research," Reuters, May 24, 2007.

93. Hartung and Goldberg, "Protecting More than Animals."

94. Reported by Heidi Prescott, as Amory was inducted into the Animal Rights Hall of Fame, Washington, D.C., July 4, 2000. Recording available from JOB Conference Recording Services, (202) 269-2000.

Notes

CHAPTER 7 The Greenies

1. Lecture by Ingrid Newkirk at New York Jivamukti Yoga School, September 7, 2001.
2. Tietz, "Bosshog."
3. Stokstad, "Global Loss of Biodiversity Harming Ocean Bounty."
4. Danielle Nierenberg, *Happier Meals: Rethinking the Global Meat Industry* (Washington, D.C.: Worldwatch Institute, 2005), p. 25, citing Rosamund Naylor et al., "Effect of Aquaculture on Global Fish Supplies," *Nature* 29 (June 2000), pp. 1017–24.
5. "How to Poison a River," *New York Times* editorial, August 19, 2005.
6. "A Malodorous Fog," *New York Times* editorial, August 7, 2005.
7. Juliet Eilperin, "In California, Agriculture Takes Center Stage in Pollution Debate," *Washington Post*, September 26, 2005.
8. Miguel Bustillo, "In San Joaquin Valley, Cows Pass Cars as Polluters," *Los Angeles Times*, August 2, 2005.
9. "Dutch 'Factory Farms' Stir Resentment in U.S.," *International Herald Tribune*, March 28, 2005.
10. Wade Rollins, "Hog Farms Try Collecting Gas, Making Energy," *News & Observer*, July 29, 2007, p. A01.
11. *An Inconvenient Truth*, directed by Davis Guggenheim, 2006.
12. H. Steinfeld, P. Gerber, T. Wassenaar, V. Castel, M. Rosales, and C. de Haan, "Livestock's Long Shadow: Environmental Issues and Options," 2006; available at www.virtualcentre.org/en/library/key_pub/longshad/A0701E00.pdf.
13. Daniele Fanelli, "Meat Is Murder on the Environment," *New Scientist*, July 18, 2007.
14. "Meat and the Planet," *New York Times* editorial, December 27, 2006.
15. Edward O. Wilson, *The Future of Life* (New York: Knopf, 2002), p. 33.
16. Joan Raymond, "Environment: Easy to Be Green," *Newsweek*, January 8, 2007.
17. S. L. Davis, "The Least Harm Principle May Require That Humans Consume a Diet Containing Large Herbivores, Not a Vegan Diet," *Journal of Agricultural and Environmental Ethics* 16, no. 4 (2003), pp. 387–94.
18. www.veganoutreach.org/enewsletter/matheny.html.
19. Marc Kaufman, "Sonar Called Likely Stranding Cause," *Washington Post*, April 28, 2006, p. A08.
20. "Beached Whales and Navy Sonar," *Christian Science Monitor* editorial, March 8, 2005.
21. www.nrdc.org/wildlife/marine/sonar.asp.
22. Ibid.
23. Allison A. Freeman, "Court Upholds Limits on Low-Frequency Sonar," *Greenwire*, July 27, 2006.
24. "Whales in the Way of Sonar," *New York Times* editorial, March 7, 2006, p. A20.
25. Associated Press, "Dolphins and Sea Lions May Get the Call to Defend Northwest Base," February 13, 2007.
26. Bruce Clifton, letter to the editor, *Seattle Times*, February 18, 2007.
27. Paul Watson, *Seal Wars* (Richmond Hill, Ont.: Firefly Books, 2003), p. 106.
28. Peter Heller, "Whale Warriors," *National Geographic Adventure*, May 2006.
29. Ibid.

30. Watson, *Seal Wars.*
31. C–99–3701-CAL. First amended complaint filed in the United States District Court for the Northern District of California, San Francisco Division, by Natural Resources Defense Council et al., Plaintiffs, vs. Carol M. Browner, Administrator of the U.S. Environmental Protection Council and the U.S. Environmental Protection Council, Defendants. Received February 8, 2000.
32. www.peta.org/feat/greenwash/pmac.html.
33. www.peta.org/feat/greenwash/wwf.html.
34. Vaa, "Reducing Wounding Losses."
35. Chris Bohjalian, *Before You Know Kindness* (New York: Shaye Areheart Books, 2004), p. 153.
36. National Shooting Sports Foundation, "What They Say About Hunting"; available at www.nssf.org/conservationvideos/pdfs/WTSAH_TG.pdf.
37. John Christoffersen, Associated Press, "Audubon Starts Deer Hunt at Greenwich Sanctuary," November 3, 2003.
38. Stephen S. Ditchkoff et al., "Wounding Rates of Whitetail Deer With Traditional Archery Equipment," 1998; available at www.sfws.auburn.edu/ditchkoff/ PDF%20publications/1998%20-%20SEAFWA.pdf.
39. John Muir, Letter to Henry Fairfield Osborn, July 16, 1904, in William Frederic Badè, *The Life and Letters of John Muir* (New York: Houghton Mifflin, 1924).
40. www.all-creatures.org/cash/cc2006-su-john.html.
41. www.washingtonwildlife.org as of March 2007.
42. Tina Lam, "PROPOSAL 3: Dove Hunting Does Not Fly: Voters Shoot Down Measure by More Than 2–1," *Detroit Free Press*, November 8, 2006.
43. Matthew Scully, "Canada's Cowardly Face," *National Post*, January 5, 2005.
44. "End the Seal Hunt," *National Post* editorial, January 19, 2005.
45. *Larry King Live*, March 3, 2006.
46. Sea Shepherd News Release, March 20, 2003; available at www.seashepherd.org/news/ media_030320_1.html.
47. Monte Hummel, O.C., President Emeritus World Wildlife Fund Canada, "A Conservation Challenge for Nunavut," Keynote Address, Nunavut Teachers' Conference Iqaluit, February 21, 2005; available at wwf.ca/AboutWWF/WhatWeDo/ConservationPrograms/RE-SOURCES/PDF/MonteNunavutSpeech.pdf.
48. tinyurl.com/yxo9jf.
49. Henry Fountain, "Squirrels That Predict the Future," *New York Times*, January 2, 2007.
50. Point made by Mike Markarian, from Fund for Animals, now with HSUS, in an unpublished letter to the editor.
51. Brian T. Murray and Judy Peet, "Bear Tally Increases and Hunt Will Widen to Federal Park Land," *Star-Ledger*, December 10, 2003; available at www.nj.com/news/ledger/jersey/ index.ssf?/base/news–5/107103925785800.xml.
52. David A. Fahrenthold, "In Md., Two Sides Take Aim Before Bear Hunt Begins," *Washington Post*, September 20, 2004.
53. John Holl, "New Jersey Sets Bear Hunt for Six Days in December," *New York Times*, November 16, 2005, p. B4.

54. Vincent Mallozzi and John Holl, "Using Guns and Honey, Hunters Take Aim at New Jersey's Bear Population," *New York Times*, December 6, 2005, p. B1.

55. Nelson Hernandez, "Girl, 8, Credited With Year's 1st Bear Kill," *Washington Post*, October 25, 2005, p. B1.

56. The *Record*'s staff, "Dead Bear Seen on Route 23 Was Victim of Car, Not Hunters," *Record*, November 23, 2003.

57. Priscilla Feral, "Deer Hunters," *Newark Advocate*, November 10, 2002.

58. Ted Williams, "Natural Allies," *Sierra Magazine*, 2000; available at www.sierraclub.org/sierra/199609/allies.asp.

CHAPTER 8 Compassion in Action

1. Taken by the Gallup Organization on 1,004 U.S. adults, April 2000, cited in *Understanding the Public Image of the U.S. Animal Protection Movement*, produced by the Humane Research Council on behalf of a coalition of animal protection groups, March 2004.

2. Kurt Brungardt, "Galloping Scared," *Vanity Fair*, November 2006.

3. Ibid.

4. "The Horse Is Saved," *Washington Times* editorial, March 30, 2007.

5. Ibid.

6. Kevin Moran, "Horse-Meat Export Fight Resumes," *Houston Chronicle*, March 18, 2007, p. B4.

7. "Stop Horsing Around," *Washington Times* editorial, July 25, 2006, p. A16.

8. "Support for Laws in California Regulating the Raising of Calves and Pigs for Food," submitted to Gene Bauston, Farm Sanctuary, by Zogby International, April 24, 2003.

9. Henry C. Jackson, "Challenge to Farm Emissions Rejected," *San Francisco Chronicle*, July 18, 2007.

10. "Swine by Us—Court Rules Against Green Groups, Lets Factory Farms off the Hook," *Grist*, July 19, 2007.

11. Lauren Etter, "Smithfield to Phase Out Crates. Big Pork Producer Yields to Activists, Customers, on Animal-Welfare Issue," *Wall Street Journal*, January 25, 2007, p. 14.

12. Brody Mullins, "Puppy Power: How Humane Society Gets the Vote Out," *Wall Street Journal*, November 7, 2006.

13. Marianne Williamson, *The Power to Change* box set, Disc 4, "The Living Temple," Hay House, 2005.

14. Scully, *Dominion*.

15. *Real Time with Bill Maher*, April 13, 2007.

16. Calum Macdonald, "Row as Charity Posters Compare Disabled to Pets," *Herald*, January 11, 2007.

17. Ibid.

18. Andrew Tyler, "Don't Follow the Herd and Give a Cow for Christmas: These Gifts Are Not a Good Thing. They Serve Only to Increase, Not Diminish, Poverty," *Independent*, November 27, 2006.

19. Personal e-mails received Decemeber 12, 2006.

20. Genesis Awards at Beverly Hilton, March 16, 2002; aired on Animal Planet, May 19, 2002.

21. Peter Singer, ed., *In Defense of Animals: The Second Wave* (Malden, Mass.: Blackwell, 2005).

22. Duane Pohlman, King 5 Seattle, May 24, 2000.

23. Paul McCartney, quoted in Sebastian Smee, "Beastly Goings On," *Weekend Australian*, March 10, 2007, p. 18.

24. Personal telephone interview with Karen Dawn in 2005 for chapter in Singer, ed., *In Defense of Animals*.

25. Conducted by the National Conference of Editorial Writers, 1994.

26. Photo by Myung J. Chun, in Steve Hymon, "Conferees Ponder: 'How Vegan Is Enough?'" *Los Angeles Times*, August 3, 2003.

27. Photo by Carlos Chavez, in Carla Hall, "Pitbulls Out of the Doghouse," *Los Angeles Times*, August 3, 2006.

28. Enrique Zaldug, "In Spain, No Olé for Bullfights," *Time*, August 6, 2007.

29. Christopher Clarey, "Surviving the Bull Market to Die in the Ring," *International Herald Tribune*, May 4, 2004.

30. Dan Noyes, KGO ABC "I-Team," September 16, 2003.

31. Personal telephone interview with Karen Dawn in 2005 for chapter in Singer, ed., *In Defense of Animals*.

32. President of Animal Liberation Victoria.

33. Michelle York, "Hen Activist Says the War on Cages Will Go On," *New York Times*, May 7, 2006, p. A40.

34. *Primetime*, May 11, 2006.

35. Steve Best and Anthony Nocella, *Terrorists or Freedom Fighters: Reflections on the Liberation of Animals* (New York: Lantern Books, 2004).

36. Jehangir S. Pocha, "Outsourcing Animal Testing. U.S. Firm Setting Up Drug-Trial Facilities in China, Where Scientists Are Plentiful but Activists Aren't," *Boston Globe*, November 25, 2006.

37. "Being Dumb with the Dumb," *Sunday Times* editorial, October 2, 2005, p. 16.

38. Philip Johnston, "Public Turns on Animal Terrorists," *Daily Telegraph*, May 29, 2006.

39. Press Association article on *Guardian* Web site, May 29, 2006.

40. Committee on Environment and Public Works, October 26, 2005; audio available online at epw.senate.gov/epwmultimedia/epw102605.ram.

41. Mark S. Blumberg, "The Animal Zealotry That Destroyed Our Lab," *Washington Post*, July 17, 2005, p. B03.

42. *60 Minutes*, "Eco-Terror's Growing Threat," November 13, 2005.

43. "An Enemy of the States," *Toronto Star*, March 13, 2006, p. A06.

44. Robert B. Cialdini, *Influence: The Psychology of Persuasion*, rev. ed. (New York: Collins, 1998).

45. La Ferla, "Uncruel Beauty."

46. Williamson, "The Living Temple."

Acknowledgments

I thank Gloria Steinem, Peter Singer, and John Coetzee for their loving guidance, support, and incomparable example.

Warm thanks to reporter Jane Velez Mitchell, and to Jinky's mom, Carole Davis—both superb activists and extraordinarily generous friends.

Guru Singh kept me focused on my mission, Brenda Shoss intermittently but effectively encouraged me to write a book (When's yours, Brenda?), and my glorious friend Nancy Hall offered unflagging love and encouragement.

I thank my brilliantly gifted friends, the artist Anthony Freda and cartoonist Dan Piraro, for their enthusiastic support and fabulous work. Also the remarkable artist Sue Coe, PCRM's Doug Hall, Linda Frost, and Dave Farley made particularly generous art contributions. And Noah Lewis's Vegetus Web site helped me collect cartoons.

Matthew Scully, Dan Matthews, Adam Weisman, Rick Bogle, Peter Muller, and Stewart David have challenged and influenced my thinking. Rolf Wicklund has been enormously helpful with Web sites and promotion, and producer Joshua Katcher has infused the atmosphere around this project with his cool warmth.

If you find yourself laughing while reading this book it will often be thanks to suggestions from the brilliant comedian Ritch Shydner. I also thank Allan Havey for some great lines and ideas.

At PETA, the amazing Ingrid Newkirk offered warm support. Alka Chandna and Shalin Gala, Debbie Leahy and RaeLeann Smith, John Machuzak and Lisa Lange, and Jessica Sandler were all fantastic. Bruce Friedrich, your loyal and constant friendship and beautiful example over the years has meant more to me than you will ever know.

HSUS's Wayne Pacelle's friendship, encouragement and belief in my work helped keep me in the movement in my early years; he taught me much. Beverly Kaskey and Leigh O'Brien have been terrific.

At Farm Sanctuary, the lovely Samantha Ragsdale has been fantastically supportive, and Chloe-Jo has been, of course, fabulous.

Thanks to my "Farmed Animal Answer Woman," Mary Finelli, and Dr. Ray Greek, for their thoughtful comments on chapters 5 and 6 respectively, to Susan Clay and Lew Regenstein for constant encouragement, and to Paul Shapiro, Jack Norris, Dan Kinburn, Chris Anderlik, Peter Muller, David Cantor, and Joe Miele for generously responding to requests for information.

Janice Blue got me into radio; none of the radio quotes would exist without her.

My thanks to my Peet's buddies Steve Godchaux, Bill Birrel, and Doug Green for brainstorming with me. (Bill, "All the World's a Cage" is my favorite chapter title.) I especially thank Pam and Chrissie Rikkers for going out of their way to help.

At HarperCollins my thoughtful editor, Katherine Nintzel, has shown true commitment to this book and sometimes saintly patience. Carrie Kania, David Roth-Ey, Calvert Morgan, Kolt Beringer, and Heather Drucker have been a wonderful team. And I thank my agent, Kirsten Neuhaus, for her keeping us on course.

If I have pointed to your work for the animals on these pages, or you have endorsed this book, or enthusiastically agreed to have your photograph included, please consider yourself wholeheartedly thanked.

Years ago I sang in the choir before Marianne Williamson's lectures at Town Hall. Those lectures brought me to a Course in Miracles, and have recharged my spirit when it ran low. I hope Marianne is happy to know she has a significant hand in any good that comes from my work.

One year, watching the Oscars and Grammies, I noted that at the Grammies everybody thanked God, while at the Oscars nobody did. No surprise that I am a musician, for I could not imagine sending in this book without giving thanks to God, as I understand him or her or it: the absolutely loving, forgiving, and ultimately all-powerful force with which, or whom, each of us can choose to align.

And finally, I thank Jim Holcomb—God only knows where I'd be without him.

Index

Index

Index

Index